Hans Bethe
and his Physics

HANS BETHE
AND HIS PHYSICS

edited by

Gerald E. Brown
SUNY, Stony Brook

Chang-Hwan Lee
Pusan National University, Korea

NEW JERSEY · LONDON · SINGAPORE · BEIJING · SHANGHAI · HONG KONG · TAIPEI · CHENNAI

Published by

World Scientific Publishing Co. Pte. Ltd.
5 Toh Tuck Link, Singapore 596224
USA office: 27 Warren Street, Suite 401-402, Hackensack, NJ 07601
UK office: 57 Shelton Street, Covent Garden, London WC2H 9HE

Library of Congress Cataloging-in-Publication Data
Hans Bethe and his physics / editors, Gerald E. Brown, Chang-Hwan Lee.
    p. cm.
  Includes bibliographical references.
  ISBN 9812566090 -- ISBN 9812566104 (pbk)
  1. Bethe, Hans A. (Hans Albrecht), 1906–2005. 2. Quantum electrodynamics.
 3. Astrophysics. I. Brown G. E. (Gerald Edward), 1926–   II. Lee, Chang-Hwan.

QC16.B46 H36 2006
530--dc22

                                                                2006045268

British Library Cataloguing-in-Publication Data
A catalogue record for this book is available from the British Library.

Cover photo courtesy of Cornell University.

Copyright © 2006 by World Scientific Publishing Co. Pte. Ltd.

All rights reserved.

Printed by Mainland Press Pte Ltd

Preface

At the age of 97, Hans Bethe asked his long-term collaborator, Gerry Brown, to explain his physics to the world. A glance at Hans Bethe's published legacy — almost eight decades of original research, hundreds of scientific papers, numerous books, countless reports spanning the key areas of twentieth-century physics — is sufficient to realise that this was no mean task. In answering Bethe's request, the editors enlisted the help of experts in the different research fields, collaborators and friends of this "last giant" of twentieth-century physics. *Hans Bethe and His Physics* is the result, and while the book's primary aim was to explain the science behind the man, the different contributions also allow the reader to take a glimpse at the man behind the science.

The book consists mostly of papers and articles written and published previously in appreciation of the work done by Hans Bethe. It is divided into four parts. Part 1 contains papers by colleagues and friends who, over shorter or longer periods of time, had the opportunity to see Hans Bethe at work at close range. In "Hans Bethe and His Physics," Gerry Brown summarises and evaluates Hans Bethe's long career as a researcher and throws light on their collaborative efforts in the last decades of the twentieth century. "My Life in Astrophysics" is a paper initially written by Hans Bethe in 2003 as a personal reminiscence of more than six decades of research in astrophysics, work that included, among others, his discovery of the CN cycle which would earn Bethe the Nobel Prize in 1967.

For the last 25 years of their collaboration, Gerry Brown and Hans Bethe spent the month of January together at some University on the west coast, often at Caltech, but sometimes at Santa Barbara or Santa Cruz. Chris Adami, in "Three Weeks with Hans Bethe" reflects on a short period when, in 1992, for the first time he joined the two professors during their annual research gathering. The piece consists of detailed notes taken by Adami at the end of each day recalling the conversations — often at mealtimes — that had taken place throughout the day. They present a rare opportunity to see Hans Bethe through the eyes of an "intellectual grandson," and they demonstrate the powerful impression Hans Bethe left with those who had the chance to learn from him in such close proximity. Jeremy Bernstein, in "Hans Bethe at *The New Yorker*", recalls the story behind his profile of Hans Bethe in *The New Yorker*, the basis of his later biographical sketch *Hans Bethe, Prophet of Energy*. Ed Salpeter's acquaintance with Hans Bethe went back to the immediate post-war years, when he first encountered his future postdoctoral adviser as a graduate student under Rudolf Peierls at Birmingham. In "My Sixty Years with Hans Bethe," Salpeter, by looking at the way Bethe impacted upon his own development as a scientist, dwells on the formative influences of Bethe as researcher and as a teacher, tracing those back to the latter's own teachers, Arnold Sommerfeld and Enrico Fermi.

Part 2 of this volume consists of the introduction and five papers reprinted from a special issue of *Physics Today*, in which guest editor Kurt Gottfried asked a number of colleagues to commemorate Hans Bethe's contribution to twentieth-century physics. After Gottfried's general appreciation of Bethe's work, Silvan Schweber writes about the period preceding the Second World War. The four other papers in this section deal with four distinct aspects of Bethe's work. The late John Bahcall and Ed Salpeter discuss Bethe's work on energy production in stars, nuclear astrophysics and neutrino physics; Freeman Dyson traces Bethe's influence on the development of quantum electrodynamics; John Negele analyses Bethe's work on the theory of nuclear matter and in particular his post-war contribution to understanding the nuclear many-body problem; and Gerry Brown presents a picture of his collaboration with Hans on supernovae and other astrophysical topics in the last quarter of the twentieth century.

Part 3 contains papers which investigate further areas of Hans Bethe's research and teaching interest. C. N. Yang and Mo-Lin Ge, in "Bethe's Hypothesis" sketch the origins and impact of what Yang himself had termed the "Bethe Ansatz." The other three papers deal with three of the key areas of Hans Bethe's scientific activity. Starting off as a solid state physicist with his first publications in 1927, Bethe moved the focus of his attention towards nuclear physics to solid-state physics in the early to mid-1930s, but he continued to make substantial contributions into the early post-war years. N. David Mermin and Neil Ashcroft look back on the first twenty years of Bethe's scientific activity investigating just how influential a solid-state physicist Hans Bethe was throughout the 1930s. Jeremy Holt and Gerry Brown, after a tour de force of the history of nuclear physics, put Hans Bethe's contributions to our understanding of the nuclear many-body problem into their historical context. Finally in this section, and appropriate as a conclusion to Hans Bethe's achievements, is "And Don't Forget the Black Holes," a paper which he co-authored with Gerry Brown and Chang-Hwan Lee shortly before his death, in 2005. As much of his work during the last quarter of a century of his life it was on astrophysics phenomenon.

The final part contains papers which go beyond a discussion of Hans Bethe's contribution to scientific progress in the twentieth century. They describe why Bethe, notwithstanding his outstanding mathematical and scientific powers, had such a huge impact on public policy. Sidney Drell, in "Shaping Public Policy," recounts the various ways in which Hans Bethe made his voice heard, be it in defence of beleaguered colleagues during the McCarthy Era, be it within the scientific community when difficult choices about the direction of research had to be made, or be it in his role as adviser to US Governments and Presidents, convincing them of the folly or wisdom of different policy options. Boris Ioffe examines Hans Bethe's stance of ways to solve the global energy problems, one of the latter's main political concerns. While Bethe had always been a staunch supporter of nuclear disarmament, he had, at the same time, advocated the peaceful use of nuclear energy, a position which he never tired to explain and which Ioffe takes up in his description of Bethe's views on nuclear power. Finally, obituaries by Richard L. Garwin and Frank von Hippel and Kurt Gottfried respectively demonstrate once more Bethe's outstanding qualities as a scientist and human being alike.

May 2006

G. E. Brown
C.-H. Lee

Contents

Preface ... v

List of Contributing Authors ... xiii

Part 1

Hans Bethe and His Physics ... 1
  Gerald E. Brown

My Life in Astrophysics ... 27
  Hans A. Bethe

Three Weeks with Hans Bethe ... 45
  Christoph Adami

Hans Bethe at *The New Yorker* ... 111
  Jeremy Bernstein

My Sixty Years with Hans Bethe ... 115
  Edwin E. Salpeter

Part 2

Hans Bethe ... 125
  Kurt Gottfried

"The Happy Thirties" ... 131
  Silvan S. Schweber

Stellar Energy Generation and Solar Neutrinos 147
John N. Bahcall and Edwin E. Salpeter

Hans Bethe and Quantum Electrodynamics 157
Freeman Dyson

Hans Bethe and the Theory of Nuclear Matter 165
John W. Negele

Hans Bethe and Astrophysical Theory 175
Gerald E. Brown

Part 3

Bethe's Hypothesis 185
Chen Ning Yang and Mo-Lin Ge

Hans Bethe's Contributions to Solid-State Physics 189
N. David Mermin and Neil W. Ashcroft

Hans Bethe and the Nuclear Many-Body Problem 201
Jeremy Holt and Gerald E. Brown

And Don't Forget the Black Holes (with Commentary) 239
Hans A. Bethe, Gerald E. Brown and Chang-Hwan Lee

Part 4

Shaping Public Policy 251
Sidney Drell

Hans Bethe and the Global Energy Problems 263
Boris Ioffe

In Memoriam: Hans Bethe 273
Richard L. Garwin and Frank von Hippel

Obituary: Hans A. Bethe 279
Kurt Gottfried

List of Publications of Hans A. Bethe 283

List of Contributing Authors

Christoph Adami, Professor of Applied Life Sciences, Keck Graduate Institute, Claremont, California and Faculty Associate at the California Institute of Technology, Pasadena, California
Christoph_Adami@kgi.edu

Neil W. Ashcroft, Horace White Emeritus Professor of Physics at Cornell University, Ithaca, New York
nwa@ccmr.cornell.edu

John N. Bahcall, Professor of Natural Sciences at the Institute for Advanced Study, Princeton, New Jersey
(passed away on 17 August 2005)

Jeremy Bernstein, Emeritus Professor of Physics at the Stevens Institute of Technology, Hoboken, New Jersey, and ex-staff writer for *The New Yorker*
jbernste@earthlink.net

Gerald E. Brown, Distinguished Professor of Physics at the State University of New York, Stony Brook, New York
gbrown@insti.physics.sunysb.edu

Sidney D. Drell, Senior Fellow at the Hoover Institution and Professor Emeritus of Theoretical Physics at the Stanford Linear Accelerator Center (SLAC), Stanford University, Stanford, California
drell@slac.stanford.edu

Freeman Dyson, Emeritus Professor of Physics at the Institute for Advanced Study, Princeton, New Jersey
dyson@ias.edu

Richard L. Garwin, IBM Fellow Emeritus at the T.J. Watson Research Center, Yorktown Heights, New York
rlg2@us.ibm.com

Kurt Gottfried, Emeritus Professor of Physics at Cornell University, Ithaca, New York
kg13@cornell.edu

Jeremy Holt, Department of Physics and Astronomy, State University of New York, Stony Brook, New York
jeholt@grad.physics.sunysb.edu

Boris Ioffe, Head of the Laboratory of Theoretical Physics at the Institute of Theoretical and Experimental Physics, Moscow, Russia
ioffe@itep.ru

Chang-Hwan Lee, Professor of Physics at Pusan National University, Pusan, Korea
clee@pusan.ac.kr

Ge-Mo Lin, Professor of Mathematics at Nankai Institute of Mathematics, Nankai University, Tianjin, China
geml@nankai.edu.cn

N. David Mermin, Horace White Professor of Physics at Cornell University, Ithaca, New York
ndm4@cornell.edu

John W. Negele, William A. Coolidge Professor of Physics at the Massachusetts Institute of Technology, Cambridge, Massachusetts
negele@lns.mit.edu

Edwin E. Salpeter, J.G. White Distinguished Professor in the Physical Sciences, Emeritus, at Cornell University, Ithaca, New York
ees12@cornell.edu

Silvan S. Schweber, Emeritus Professor of Physics and the History of Ideas at Brandeis University, Waltham, Massachusetts
schweber@brandeis.edu

Frank von Hippel, Professor of Public and International Affairs and Co-Director, Program on Science and Global Security, Woodrow Wilson School, Princeton University, Princeton, New Jersey
fvhippel@princeton.edu

Chen Ning Yang, Albert Einstein Emeritus Professor of Physics at the State University of New York, Stony Brook, New York and Professor of Physics, Tsinghua University, Beijing, China
yang@castu.tsinghua.edu.cn

Hans Bethe and His Physics

Gerald E. Brown

In 2003 I put together, with Chang-Hwan Lee, a book on *Formation and Evolution of Black Holes in the Galaxy*.[1] This collection represented papers that Hans and I had published together from 1979 through to 2002, some of them written with other collaborators, especially with C.-H. Lee in the later years. Seven years earlier, Hans had already published his own *Selected Works*.[2] Hans was 97 years old when *Formation and Evolution of Black Holes in the Galaxy* came out. Each paper in this volume was accompanied by a commentary, giving some history of how we came up with the ideas we used. Hans read carefully and critically every commentary, suggesting modifications and then saying that he could not improve further on them.

Then Hans asked me to explain his physics to the world. I began by writing a draft of "Hans Bethe and the Nuclear Many-Body Problem,"[a] summarizing work he had begun in 1955, the year up to which his official biographer Sam Schweber had agreed to write his biography. In fact, I had followed this work since 1955 while working with Rudi Peierls in Birmingham, Hans' closest friend. Hans was on sabbatical in Cambridge that year. (I worked with Peierls for 10 years which prepared me for working with Hans.) I sent a draft of my paper to a close friend, a great expert in the nuclear many body problem, who replied "Do you really want to do that?" and was highly critical of the manuscript, basically saying that only nuclear physicists, those working or who had worked in the many-body problem, would be able to understand my article, which was badly written.

---

[a]See p. 201, this volume.

So I decided to find experts, those especially knowledgeable about each of the different subjects or topics Hans had worked on, to write reviews of his work and further applications. Biographers would then have materials to work with.

In particular, Hans had done nothing further after publishing it on the subject of his paper "On the Theory of Metals, I. Eigenvalues and Eigenfunctions of a Linear Chain of Atoms."[3] Yet, at the Celebration for his 60 years at Cornell, this paper had the widest application of any of his publications in the talks. Hans simply had not followed the many applications of this paper in statistical mechanics and related subjects. I knew that my colleague C. N. Yang had invented the name "Bethe Ansatz" and include his review in this volume. In fact I phoned Hans the (Sunday) morning of March 6, 2005, to talk about the review paper "Evolution and Merging of Binaries with Complex Objects," which he and I and Chang-Hwan Lee were writing for his Centenary Volume of *Physics Reports*,[4,b] discussing some points with him, and then told him that Yang was going to write about the "Bethe Ansatz," which greatly pleased him. That evening Hans died peacefully while eating supper.

The problem of finding experts on the problems Hans solved turned out to be not so difficult, except that he worked on so many problems in so many areas that I had to find a lot of them. I tried also to put personal touches that would portray Hans' amazing qualities of reason and judgment.

I was asked, together with Frank von Hippel, to write the Memoirs of Hans for the British Royal Society. These memoirs generally include biographical sketches of the early life of the subject. I remembered that Jeremy Bernstein had written articles about Hans' life after a series of interviews beginning November 1977, interviewing Hans over a period of two years for *The New Yorker* and that these are published in his book, *Hans Bethe, Prophet of Energy*.[5] Hans was obviously very pleased with the result and had sent me a copy inscribed "To my closest collaborator, Gerry Brown; Hans Bethe." I began referring to the book for Hans' early life, and then found that I was actually copying it, aside from inserting small additions from the evenings spread over the 25 or so years Hans and I worked together. (We spent Januaries together, on the west coast, in Santa Barbara, Santa Cruz and most often in Caltech. Until the last years Rose stayed at home to take care of her parents and my wife Betty stayed home to look after the children, so I did the cooking and housecleaning. Two different years Hans was our

---

[b]A popularized version of some of this is included as "And Don't Forget the Black Holes" in this volume.

boarder in Copenhagen in the spring.) The atmosphere of the many evenings we spent together is recreated in this volume by Chris Adami's *Three Weeks with Hans Bethe*.

I talked with Jeremy Bernstein on the phone, explaining my dilemma of chiefly using his words, and he answered "Why don't you just copy my book?" Aside from my small changes I noted above, in Section 1 of the following, I have done just that. I am very grateful to Jeremy for permission. I also include the story he sent me about how his articles would very nearly not have been published in *The New Yorker* which would have robbed the world of Jeremy's very accurate picture of how Hans worked and what he did.

1. German Beginnings

Hans Bethe was born in Strasbourg, in Alsace-Lorraine on July 2, 1906. His father was a physiologist, Privatdozent at the University of Strasbourg. He had come from Stettin, in what was then the northern Prussian province of Pomerania, now part of Poland.

Hans' father's family was Protestant and through history some members were Protestant ministers and some were schoolteachers. Hans' grandfather was a medical doctor.

Hans Bethe's father reached his position in a somewhat unorthodox way, as a late developer, having failed four times in his school years to be promoted. After he graduated from the Gymnasium at age twenty he studied intensively and finished his Ph.D. in zoology at the University of Munich. Then he went to Strasbourg to do physiology, taking a medical degree on the way.

The family of Hans' mother was Jewish. Her father was professor at the University of Strasbourg. His specialty was ear, nose and throat diseases. For a Jew to get a professorship at that time was quite exceptional.

In 1912 the Bethes moved from Strasbourg to Kiel where Hans' father became chairman of the Physiology Department at the University of Kiel. In Kiel, Hans, who was a rather frail child, was one of a small group of children studying under a private tutor rather than going to a regular school. His life was spent almost entirely with grownups, parents and relatives. There was a lot of solitary contemplation.

Hans' father did talk to him about scientific matters. He knew some mathematics, mainly algebra, and in his work he used a slide rule. He had learned some calculus, but could not apply it in calculations.

In 1915 the Bethes moved on to Frankfurt where Bethe's father had been invited to start a Department of Physiology at the University of Frankfurt, which had been founded only one year before. Indeed, the University of Frankfurt claims Hans Bethe as their native son and one can still pass the house where he grew up on Kettenhofweg, in walking from the Musikstrasse to the old university. A square in the new university is named after Hans Bethe.

In Frankfurt Hans began to attend the Gymnasium. He easily passed the entrance exam. This was a new experience for him in that he made friends of his age. He was nine years old when he began. The gymnasium had nine years altogether. In Hans' last few years he took some elective physics courses. They were like US college freshman physics courses and included lots of mechanics. The students used calculus to calculate trajectories, for example.

When Hans was 18 he published his first paper with his father and one of his colleagues on the theory of dialysis.[6] He used calculus to calculate flows in bodily fluids, the flows being driven by chemical potentials. Amusingly he used the same equations late in his life to calculate the flow of degenerate neutrinos formed in the interior of large collapsing stars, in our collaboration on the theory of supernovae.

Hans entered the University of Frankfurt in 1924. He took courses in physics, chemistry and mathematics. One of the associate professors, Walther Gerlach, was a young man of terrific enthusiasm. In his course in modern physics he discussed electrons, spectroscopy, etc. This was very, very interesting to Hans Bethe. Hans also took a course in the modern theory of numbers from Carl Ludwig Siegel, a distinguished mathematician. Gerlach left after Hans' first year and was replaced by a spectroscopist Karl Meissner. He told Hans emphatically that he must not stay in Frankfurt but should go to a place with better theoretical physics. (It was clear to Hans that he would not be good as an experimentalist.) Hans loved mathematics, but he wanted to apply it to physics.

The quantum theory in 1924 was in a state of transition. Niels Bohr, the great Danish physicist, had invented the old quantum theory, in which electrons were allowed orbits with only integer numbers of de Broglie wavelengths and the frequencies of emitted light were restricted to the differences in energies of these orbits. There was no real understanding of what stabilized these orbits against radiating away energy as they whirled around in circles or ellipses and in classical description would have collapsed into the nucleus.

In 1926 the quantum theory of wave mechanics was invented by Erwin Schrödinger and Werner Heisenberg, although it took some time to show that the two were the same theory in different forms.

After Hans Bethe had been in Frankfurt for two years Meissner strongly advised him to leave for Munich. He not only told Hans to leave Frankfurt but specifically to work with Sommerfeld, the professor of theoretical physics at the University of Munich. In 1926 Arnold Sommerfeld was the most influential physics teacher in the world, his recent prize pupils having been Wolfgang Pauli and Werner Heisenberg, who by 1923 had both left Munich. Heisenberg received the Nobel Prize in 1932 and Pauli in 1945. Before then Sommerfeld had had among his students Peter Debye and Max von Laue, who both became Nobel Prize winners.

Sommerfeld worked in every area of theoretical physics and his lectures formed one of the best introductions to many branches of physics. There is no doubt but that Sommerfeld influenced Hans Bethe to do the same. In May of 1926 Hans presented to Sommerfeld letters from his father who had met him by chance somewhere during the First World War and a letter from Meissner. Probably the latter counted more. Sommerfeld said "all right, you are very young, but come to my seminar."

Sommerfeld had at that time an official assistant, Gregor Wentzel, who had an office next to him. There was another room of about the same size which was library and home to the students. The foreign postdoctorates and German graduate students sat in this room. Important for Hans Bethe's later life, Rudolf Peierls, who was one year younger than Hans, was one of these students. Lifelong collaboration between Bethe and Peierls began (which was also of great benefit to me because I worked for 10 years with Rudi Peierls and more than twenty-five years with Hans Bethe).

Sommerfeld was at his greatest as a teacher. The first course Hans attended was in differential equations of physics, which was his specialty. The seminar was once a week and that was where the student learned to analyze new material and to enable others to understand it.

Schroedinger's wave function method, as compared with Heisenberg's matrix formulation, was highly preferable to Sommerfeld, who adapted his polynomial method of solution to it. Hans Bethe came at just the time of the new physics, unencumbered by old concepts and theories. He remembered what he worked out with Sommerfeld for the remainder of his life, and was able to access this knowledge for other problems.

Hans Bethe's thesis, suggested by Sommerfeld, was a theoretical analysis of the Davisson and Germer[7] diffraction of electrons by crystals. Since electrons have a wave property, as evidenced by their de Broglie wavelength, they diffract in a similar way to the X-rays investigated some years before by Max von Laue. Sommerfeld suggested that Bethe look at a paper by Paul Ewald, who had dealt with the diffraction of X-rays by crystals, also at Sommerfeld's institute. So Hans started out by making a more or less direct translation of Ewald's thesis from X-ray to electron and found out that that worked very well.

In 1928 having passed his doctoral exams, Hans began looking for a job. Although positions were few, Hans received a teaching offer from Professor Madelung, who had been one of his teachers in Frankfurt, and he accepted it. Hans was to do research and, together with an assistant in experimental physics, teach the elementary laboratory. The atmosphere had, however, not changed in response to the new wave mechanics, so Hans settled down to reading and learned group theory. His paper on the splitting of wave functions in crystals[8] is widely used, especially by physical chemists. The problem was more tractable than his thesis problem, which involved strong interactions of the electrons with the material, and the issues more cleanly settled by symmetries.

The environment in Frankfurt had not reacted to the new quantum theory, so it was lucky for Hans that after half a year there he was asked by Ewald to come to Stuttgart as his assistant. Paul Ewald was the professor of theoretical physics at the Technical College of Stuttgart.

Hans was absorbed with the Ewald family, later in a more literal sense when he married Paul Ewald's daughter Rose. He was asked to lecture twice a week on the new quantum mechanics to Ewald and all of his assistants, also visitors, who came from all over the world to study with Ewald. One of the most useful visits was by Douglas Hartree who lectured about his self-consistent method for calculating atomic states in large atoms. This method begins from an Ansatz for all of the individual levels, calculated in a Thomas-Fermi approximation to the screened Coulomb potential from the nucleus and then obtains wave functions which are products of single-particle wave functions by iteration. These general ideas of how to proceed in a self-consistent fashion strongly influenced Hans' thinking about the nuclear many-body problem which he attacked only much later, beginning in 1955. A salient characteristic of Hans' life was that he was at the right place at the right time, although when he became older, during his time in America, he developed the "right place" through his own work.

In the midst of Hans' happy situation in Stuttgart, Sommerfeld returned from an around the world trip. Although he had done nothing to help Hans find his first positions, he wrote a postcard to Ewald saying, in effect, "Bethe is my student. Send him back to me immediately." The two professors fought this out between them and, of course, Sommerfeld won. Sommerfeld arranged that Hans could become a Privatdozent the following spring, which was unusually early, and got him a German national fellowship.

In the winter of 1929 Hans wrote what he considers to be his best paper (although I believe several others to have been more important) on the stopping power of charged particles as they go through matter.[9] Niels Bohr had calculated the energy losses of the charged particles by exciting the electrons in atoms as they pass in 1913 and Max Born had formulated collision theory in wave mechanics, but putting the two together gave what looked like a complicated expression. Hans' great simplification was to work in momentum space where, folding operations in configuration (x, y, z) space of the various quantities are replaced by products, the Fourier transform of the Coulomb potential being particularly very simple, $4\pi e^2/q^2$ where q is the momentum.

Bethe's trick is now standard, and one forgets that someone had to invent it. In experiments one now has to sort out the different kinds of particles which emerge and the measurement of the ionization they produce in a gas, which Bethe's method makes possible, is very useful in doing this.

Bethe's paper was submitted to the university in order for him to become a Privatdozent.

In 1930 and 1931 Hans received a Rockefeller Foundation fellowship. He used it to first visit Cambridge University in England and then Rome where he worked at the University of Rome with Enrico Fermi.

Hans found that the British had a much healthier attitude toward life than the Germans. One could talk with them about politics, philosophy or anything else in a reasonable way. He used the time in Cambridge to deepen his understanding of quantum mechanics, especially from Dirac's book.

Dirac lectured from his book in a literal sense, essentially reading it aloud. (He was still doing that in 1950 when I attended one of his lectures in Cambridge.) Hans found that he could just as well stay in his "digs" and read the book as go to Dirac's lectures.

The year with Fermi was particularly useful for Hans and shaped his way of research for the rest of his life. Fermi taught him how to approximate. From Fermi he learned to look at the problem qualitatively first, and understand the problem physically before putting lots of formulas on paper. Sommerfeld never could have done that and never would. Sommerfeld's method was to begin by inserting the data of a problem into an appropriate mathematical equation and solving the equation quantitatively according to the strictest mathematical formalism for those specific data. Under his tutelage one never looked at the problem first with the view of finding the easiest way to solve it. It was from Fermi that Hans learned to do this. For Fermi the mathematical solution was more for him a confirmation of his understanding of a problem than the basis for its solution.

Although Fermi's main interest during the period of Hans' visit was low-energy neutron scattering, he and Hans coauthored a paper[10] comparing three methods of treating the relativistic electron-electron interaction unifying electromagnetic quantum theory with gravity. But in general, although Hans took an interest in the experiments he worked on his own. He worked out and published the solution of the linear chain during that period,[3] introducing what C. N. Yang named "The Bethe Ansatz." This work probably has had the most influence over a wider variety of fields of any of Hans' works. It is reviewed by C. N. Yang later in this volume.

Hans told me that Fermi had never read the paper, and Hans himself did not later use that paper, in spite of the many applications by other research workers, after publishing it.

In 1932 Hans, in collaboration with Sommerfeld, wrote one of his three great review articles "Elektronentheorie der Metalle."[11] In fact, Sommerfeld wrote the first chapter and Hans wrote the rest of the book. Quantum mechanics had greatly simplified many aspects of solids, such as their low specific heat. (Introduction of a small amount of heat into metals is sufficient to raise their temperature a lot and this was easily explained in the new quantum mechanics.) Although Hans wrote the major part of the book in less than a year, he was at the same time learning nuclear physics which he was eager to get into.

In the summer of 1932 Hans got an offer from Tübingen of a position similar to an American assistant professorship. Hans Geiger, the inventor of the Geiger counter, was the professor of experimental physics. Whereas Hans had a good time scientifically in Tübingen the general atmosphere became very bad, many students wearing brown shirts and swastika armbands of the National Socialist (NAZI) organization.

The Reichstag fire occurred on February 27, 1933 and soon Hitler took over Germany. On April 4, 1933 the first racial laws were promulgated. It was clear that according to the laws Hans could not hold a university job because two of his grandparents were Jewish.

Hans wrote to Geiger, who had been very friendly to him. Geiger wrote back a cold letter saying that it would be necessary to dispense with Hans' further services. Hans wrote to Sommerfeld, who immediately replied "You are most welcome here. I will have your fellowship again for you. Just come back." In 1933, although relatively secure under Sommerfeld's protection at that time, it became clear to him that he would have to leave Germany.

In 1933 Hans fortunately received an offer of a temporary lectureship in Manchester University in England, one of the many positions that the British under the leadership of Rutherford organized to provide temporary positions for Germans who had been fired under the new racial laws. Hans was reunited with his close friend Rudi Peierls (later Sir Rudolf Peierls) and was the first of many boarders in the Peierls house, (of which I was one later). In our many conversations about Hans' life, it was clear that he considered the year 1933–1934 as his most productive, although I would not identify any of the problems he solved as important as the three works: "Bethe Ansatz," energy production in stars, and the Lamb shift, any one of which was worthy of the Nobel Prize. Working with Rudi Peierls was highly enjoyable for Hans, and they got along very well together.

In the autumn of 1933 Bethe and Peierls made a visit to Rutherford's Cavendish Laboratory in Cambridge, traveling by train. Chadwick acquainted them with an experiment carried out with a bright young graduate student Maurice Goldhaber on the photodisintegration of the deuteron (then called diplon). Chadwick challenged them to work out the theory of this reaction. Luckily for their theoretical work, trains took a long time to go cross country in England at that time, about 4 hours from Cambridge to Manchester. Bethe and Peierls had a solution to the problem by the time they reached Manchester (1935).

Bethe and Peierls also wrote a paper on neutron-proton scattering that year,[12] but underestimated the cross section substantially because they did not know of the virtual state that Wigner later developed. At the time of this work and of the photodisintegration of the deuteron, the main characteristics of the nucleon-nucleon interaction, that it was short-ranged with effective radius about 1 fm (10^{-15} meter) and quite deep (~ 30 MeV) were already known from scaling arguments involving a number of nuclei.

Hans also wrote the paper "On the Stopping of Fast Particles and on the Creation of Positive Electrons" with W. Heitler[13] during his stay in Manchester.

Hans had known Nevill Mott, who was currently working in Bristol, during his earlier visit to England. Hans gave a talk there and intimated he would love to come there. A few weeks later Mott offered him a fellowship at Bristol for a year, but in the summer of 1934 Hans got a cable out of the blue from Cornell offering him an acting assistant professorship, with the prospect that it might be made permanent. Hans, at age 28, moved to Cornell where he worked the rest of his life.

2. America, the First Years

The Bethe Bible

In the "Bethe Bible," three articles in the *Reviews of Modern Physics*,[14–16] Hans provided, at a time when there were no comprehensive textbooks, a complete coverage of nuclear physics. Indeed, my early background came from these articles and I can attest to their great pedagogical value.

Hans was always impressed, in thinking about the nuclear many body problem, by Hartree's self-consistent fields in atoms. In fact, Hartree visited Tübingen in 1929 where Hans had his first job as assistant to Paul Ewald. Already in the "Bethe Bible" in the paper with Robert Bacher, Hans showed that the shell model worked well for the closed shell nuclei ^{16}O and ^{40}Ca. But the problem of ^{208}Pb had to wait to be solved until 1949, a whole 13 years later, by the introduction of the spin-orbit term.[17,18]

The Washington conference

Every spring a small conference was held in Washington, sponsored jointly by the Carnegie Institution and George Washington University. George Gamow and Edward Teller usually suggested the subject of the meeting. In 1938 they suggested energy production in stars. Teller convinced Hans Bethe that he should go, although Bethe at the time was working on quantumelectrodynamics, and loath to attend.

The conference turned out to have one really important piece of information. Bengt Strömgren, a well known Scandinavian astrophysicist, reported that the central temperature of the sun was now estimated as 15 million degrees, not Eddington's 40. This change came from assuming the sun to be composed of 75% hydrogen and 25% helium, rather than assuming it to have the same chemical composition as the earth. The lower atomic weight of the revised mix lowered the temperature.

The lower temperature meant that the reactions calculated in Bethe and Critchfield[19] correctly predicted the luminosity of the sun. So Bethe and Critchfield had a theory of energy production by the sun that was immediately accepted by the conference.

This left unresolved the question of energy production in more massive stars. From observations one could show that the core temperatures increase slowly with increasing mass, but luminosity increases very rapidly. The proton-proton reaction could not predict this, as the rate of the reaction increases fairly slowly as the core temperature rises.

The other key question explored at the conference was how to build elements heavier than helium. The major problem was that no nucleus $A = 5$ exists nor does any of weight 8. Both immediately disintegrate, as shown by laboratory experiments. Elements of weight 6 or 7 in a proton sea will quickly decay into two helium atoms in reactions similar to

$$^{7}\text{Li} + \text{H} \rightarrow 2\ ^{4}\text{He}. \tag{1}$$

Ed Salpeter, a colleague of Hans at Cornell, eventually solved the problem of getting further. He noted that the $\alpha + \alpha$ reaction goes to an excited state of ^{8}Be,

$$^{4}\text{He} + ^{4}\text{He} \rightarrow ^{\star}\text{Be}, \tag{2}$$

which lasts long enough that $^{4}\text{He} + ^{\star}\text{Be} \rightarrow ^{12}\text{C}$. In fact, this is really a three body process of $3\ ^{4}\text{He} \rightarrow ^{12}\text{C}$, whereas ^{12}C is burned chiefly in a two-body process

$$^{12}\text{C} + ^{4}\text{He} \rightarrow ^{16}\text{O} + \gamma. \tag{3}$$

The fact that ^{12}C is removed by a two-body process has major implications for black-hole formation as end product of stellar burning.[20]

The carbon cycle

Because the observed energy production increased faster than the proton-proton reaction could explain, there had to be another reaction, and it had to involve heavier nuclei.

Hans went back to his office at Cornell University and developed the 6 step cycle in which carbon and oxygen act as catalysts in producing a ^{4}He nucleus from the 4 hydrogen atoms. He did this in a few weeks, simply looking through the possible reactions which had been measured by Willy Fowler and his collaborators at Caltech.

$$^{12}\text{C} + \text{H} \rightarrow ^{13}\text{N} + \gamma \tag{4}$$

$$^{13}\text{N} \rightarrow ^{13}\text{C} + e^{+} + \gamma \tag{5}$$

$$^{13}\text{C} + \text{H} \rightarrow {}^{14}\text{N} + \gamma \qquad (6)$$

$$^{14}\text{N} + \text{H} \rightarrow {}^{15}\text{O} + \gamma \qquad (7)$$

$$^{15}\text{O} \rightarrow {}^{15}\text{N} + e^+ + \gamma \qquad (8)$$

$$^{15}\text{N} + \text{H} \rightarrow {}^{12}\text{C} + {}^4\text{He} \qquad (9)$$

All of these nuclei were well known in the laboratory. Note that at the end, the starting ^{12}C nucleus is recovered, and four protons have been combined into an α-particle, just as in the chain starting with the proton-proton reaction, but through a totally different mechanism. The ^{12}C nucleus serves as a nuclear catalyst, so that relatively few carbon nuclei are needed to allow frequent occurrence of the reaction.

Originally Hans had submitted the paper to the *Physical Review*. But then Bob Marshak told him of the A. Cressy Morrison prize of the New York Academy of Sciences. He requested the paper back from *Phys. Rev.* and submitted it to them, although it was later published in the *Physical Review*.[21] He won the prize of $500. He used half of it for the Nazi officials in order to buy his Jewish mother's way out of Germany, and the other $200 was sufficient to bring both his mother and her furniture to the US. He gave Marshak a finder's fee of $50. (Marshak's thesis with Hans on white dwarfs was trail breaking.)

There is a fascinating scientific epilogue to this tale. To understand it we must expand somewhat on the solar *p*–*p* fusion process. As we have indicated, the initial fusion reaction — the one whose probability Bethe computed — is:

$$p + p \rightarrow d + e^+ + \nu. \qquad (10)$$

But once formed the deuteron (d) rapidly fuses with another proton to make a light isotope of helium (^{3}He). Symbolically:

$$d + p \rightarrow {}^3\text{He} + \gamma. \qquad (11)$$

Now the helium can collide and there are two possibilities:

$$^3\text{He} + {}^3\text{He} \rightarrow {}^4\text{He} + p + p \qquad (12)$$

or

$$^3\text{He} + {}^4\text{He} \rightarrow {}^7\text{Be} + \gamma. \qquad (13)$$

Here ^{7}Be stands for beryllium. But now the beryllium can collide and an isotope of boron (^{8}B) can be manufactured:

$$^7\text{Be} + p \to {}^8\text{B} + \gamma. \tag{14}$$

(It might seem, by the way, that this can continue indefinitely, producing all the elements in the periodic table. As it turns out, this does not work, and the heavy elements, heavier than iron, are produced in violent processes involving the collapse of old stars, with the resultant production of nova or supernova explosions.) Boron now decays in the process

$$^8\text{B} \to {}^4\text{He} + {}^4\text{He} + e^+ + \nu \tag{15}$$

in which one of the final products is a characteristic neutrino of exceptionally high energy: 10 MeV.

All neutrinos interact very weakly with matter, and these solar neutrinos escape from the sun in about three seconds. It takes the more strongly interacting gamma rays about a million years to get from the interior to the surface of the sun. But these high-energy neutrinos should be detectable here on earth. The basic identifying technique was worked out by the Brookhaven physicist Raymond Davis Jr., who has been trying to detect these neutrinos since 1968. Davis noticed that if these high-energy neutrinos impinged on chlorine they could convert the chlorine into argon by the process:

$$\nu + {}^{37}\text{Cl} \to {}^{37}\text{Ar} + e^-. \tag{16}$$

Chlorine is a good target choice since it can be stored, in large quantities, in liquids like carbon tetrachloride, a commonly used cleaning fluid. What Davis did was to place an enormous tank of the stuff, some four hundred thousand litres, nearly a mile beneath the surface of the earth (which shields it against cosmic rays) in the Homestake gold mine located at Lead, South Dakota. Periodically, the tank was flushed to see if any of the chlorine has been converted into argon by the high-energy neutrinos. For many years, Davis was not able to detect any neutrinos — something that caused great consternation among astrophysicists. Later he found neutrinos, but only about 1/3 of those predicted by John Bahcall from the standard model of the sun. Hans continued his interest in the solar neutrino problem until it was solved towards the end of his life by experiments in the Sudbury Neutrino Observatory (SNO) in 2001. I asked Hans during dinner together one evening in Santa Barbara in January 1999 how long he wanted to live. He said, until SNO experiments had sorted out the neutrino problem. The resolution of the problem is described by John Bahcall and Edwin Salpeter in their article "Stellar Energy Generation and Solar Neutrinos," in *Physics Today* and reprinted in this volume. The SNO experiment showed that the Sun's total output of neutrinos of all types was in excellent agreement with the solar model prediction for the production of electron neutrinos in the core of the sun.

On September 14, 1939, Bethe and Rose Ewald, the daughter of his Stuttgart professor Paul Ewald, were married in New Rochelle, New York. "In 1935, when I got my visa to emigrate to America, I visited the Ewalds," Bethe told me. "And Rose, who was then seventeen, was present when I told Mrs. Ewald that I was going to America. Rose said, "Why don't you take me along?" I didn't take her seriously, but two years later, when I was giving a lecture at Duke University, there she was. She had emigrated in 1936, and now she had a job as a housekeeper in Durham and was going to Duke part time. Rose had found out that I was coming to Duke, and at a banquet after the lecture she talked to me about trying to find a position for her father, who was also attempting to emigrate." He did manage to emigrate to England, in 1937, with his entire family. "Rose was then twenty, and I fell in love with her. I proposed to her in a letter as soon as I got home, but it took us two more years and a trip back to Europe on her part before we were married. In order to get a marriage license, we had to invent a domicile for her, since she had no fixed residence while going to college, and we picked New Rochelle, because it was the residence of the mathematician Richard Courant, an old family friend, who had helped Rose to immigrate and to find jobs before she went to college full time. We were married by a judge who recited the marriage ceremony in its briefest possible form — we thought it was a rehearsal. But he said now you are married, and so we are still. I handed him ten dollars, which he apparently considered more than his normal fee. After that, he was much friendlier. Richard Courant and his daughter Gertrude were there, along with Rose's housemother from Smith College, to which she had transferred in 1937, and the Tellers. Edward Teller, of course, was one of the two physicists at George Washington University who arranged the meeting that led to my work on the source of the energy of stars. In the years since we had both come to the United States, I had spent many happy hours at their home, in Washington, mostly in connection with scientific meetings. Edward and I had endless innumerable discussions on scientific problems. Later, Rose and I went for summer trips to the mountains with the Tellers. They were our best friends in this country." The friendship was one that would be severely tested in the years ahead.

The A-bomb and the H-bomb

Richard Rhodes wrote two excellent books: *The Making of the Atomic Bomb*,[22] and *Dark Sun: The Making of the Hydrogen Bomb*[23] which detail Hans' role as head of the theory group under Oppenheimer in Los Alamos where the atomic bomb was made. I cannot add to them.

My only remark is that as more and more details have come out about the bomb, I also realized more and more why Hans had phenomena like black body radiation "at the ready" in his arsenal, which was unleashed on the collapse and explosion of large stars. He had already developed the techniques for the bomb. Hans had no problem, however, in separating what was classified and should be kept to himself and what could be freely used (most of the material).

3. After the War

The Lamb shift

Here I can only summarize Freeman Dyson's elegant and masterful article "Hans Bethe and Quantum Electrodynamics" which was published in a special issue of *Physics Today* reprinted in this volume.

"From June 2 to June 4, 1947, a carefully selected group of distinguished physicists assembled at Shelter Island, a small and secluded spot near the eastern tip of Long Island." The main subject of discussion was the experiment of Lamb and Retherford who found a small deviation from the observed fine structure in hydrogen from that predicted by the Dirac theory. Many people at the conference suggested that the deviation results from the quantum fluctuation of the electromagnetic field acting on the electron in the atom. The effect of quantum fluctuations would be to give the electron an additional energy which was called the self energy, now Lamb shift.

Several, if not many, of the attendees at the meeting went home with the idea of sitting down, possibly with a graduate student or students, to make a calculation of this self energy during the next year or so. For example, Viki Weisskopf had, as student, Bruce French working on this problem. Almost all of those involved thought that a covariant relativistic calculation would be necessary. Finally it was, but only to get the last few percent of the shift.

After the meeting was over Hans Bethe traveled by train from New York to Schenectady, a distance of 75 miles. Luckily he went to Schenectady to consult for General Electric because the track ride was much smoother than the train tracks to his home, Ithaca. On the train he finished a calculation of the Lamb shift for a real electron. The value he found was 1040 megahertz, a result agreeing pretty well with Lamb's experiment. This paper was a turning-point in the history of physics. The experts began to think that the existing theory of QED was physically correct, the infinities being subtracted correctly by the renormalization procedure.

How did it happen that Hans Bethe was the one who broke the impasse? First he realized, and he was probably the only one at the meeting to do so, that the reaction of the electron to electromagnetic quantum fluctuations was mainly a nonrelativistic process and could be calculated using ordinary nonrelativistic quantum mechanics. Only Bethe had the courage to plunge ahead with a calculation using old-fashioned nonrelativistic quantum mechanics.

The "H. A. Bethe way"[c] when confronted by any problem was to sit down with his stack of paper and fountain pen and his slide rule and calculate the problem in the most obvious way. Most importantly, not to be deterred by anyone who tells you — as most colleagues will — that the situation is much more complicated than you think. One can always change to a more complicated procedure, but it is amazing how effective the H. A. Bethe way is.

Certainly the Lamb shift calculation was more important than the energy production in stars that Hans got the Nobel Prize for, and probably the "Bethe Ansatz" in his model of the linear chain, which is reviewed by C. N. Yang in this volume. It has had wider impact on theoretical physics, but the Nobel committee felt that the work for which he did receive the Nobel prize was preferable because of its connection with astronomy.[24]

The nuclear many-body problem

Schweber[25] wrote "Between 1955 and the mid-1970's, Bethe's activities in pure science were concentrated in the study of nuclear matter. Although these researches were consistently of high quality, they cannot be characterized as outstanding. The pressure of other activities — membership on PSAC and its subcommittees, consulting for AVCO, GE, and other industrial firms and crises within the home help explain the character of his scientific production during the period."

---

[c]I do not have a copyright on the "H. A. Bethe way" but I seem to be one of the few to use it.

While it is true that Hans, during this period, did not seize upon a major problem, sit down and solve it, as he had typically done before, the above judgment of his work during this period, which he undoubtedly gave to Schweber, is unnecessarily negative. This work involved the use of the G-matrix theory developed by Keith Brueckner in order to "tame" the extremely strong short-range repulsions entering into the nucleon-nucleon interaction. In fact, although these repulsions are difficult to handle mathematically, all they do is to keep the two interacting particles apart. My long-term collaborator Tom Kuo and I adapted Hans' work in the infinite system to obtain effective interactions to use in finite nuclei. We experienced a great deal of negative criticism of our work for nearly 40 years. The most negative was from my colleague Murph Goldberger at Princeton, where I was professor during 1964–1968. He turned Churchill's war commentary around to apply to those people who used Brueckner theory in the nuclear many-body problem as "Never have so many contributed so little to so few."

Yet here we are today with an interaction called $V_{\text{low }k}$ which includes all measured data from nucleon-nucleon scattering, and none from momentum regions that have not been experimentally investigated. The interaction $V_{\text{low }k}$ is widely used to calculate nuclear spectra, properties of nuclear matter, etc., etc. This procedure, or similar ones, are the only way to work out the structure of complex nuclei. This $V_{\text{low }k}$ is purely and simply the $v_l(r) + v_l \frac{Q}{e} v_l$ of Bethe et al.,[26] taken over from the work of Moszkowski and Scott.[27] Detailed discussion of $V_{\text{low }k}$ is given in this volume, in "Hans Bethe and the Nuclear Many-Body Problem" by Jeremy Holt and G. E. Brown.

One should add that in this work Hans trained excellent students, Jeffrey Goldstone, David Thouless, Phil Siemens and John Negele, whose theses were on the nuclear many-body problem. This subject gave plenty of problems to train graduate students in theoretical physics and this may have entered into his considerations of choice of research subjects. All of the students involved with them have gone on to other subjects of research since then.

Equation of state in the gravitational collapse of stars

When, together with Chang-Hwan Lee, I put together our work starting in 1978 *Formation and Evolution of Black Holes in the Galaxy*[1] I added sprightly commentaries to all of the chapters, each one consisting of a paper we wrote together, often with other authors. The great thing about these commentaries is that, although they were mostly written by me, I passed them all by Hans and he reworded them in many cases. In any case, he had his say (his opinion was never in doubt) so the commentaries as well as papers are really joint work with him. What I was amazed about in assembling the work was how little I had to modify our conclusions of the papers in the commentaries. Hans was the unique collaborator that when a paper was done it was done. Following papers were built upon it. We didn't often find mistakes or other reasons for modifying earlier papers.

In addition to this volume I wrote a short article "Hans Bethe and the Theory of Supernovae" for the special issue of *Physics Today* which is reprinted in this volume.

Freeman Dyson wrote in *Science* magazine[28] "Hans Bethe was the supreme problem solver of the past century. He was not a deep thinker like Heisenberg and Dirac, who laid the foundation of modern physics in the 1920's. But he took their theories and made them into practical tools for understanding the behavior of atoms, stars and everything in between." I began Hans' obituary in *Nuclear Physics*[29] with my friend Freeman's quotation, because I was too close to Hans, too impressed with Hans, to write an objective evaluation of his place in the past century.

What I saw was a very human collaborator, the closest of close friends, who started his research of the day with a stack of white paper on the upper left-hand corner of his desk, fountain pen in hand. Then he would begin working out whatever problem he had planned to do, remembering all constants, and would fill the white sheets at a nearly constant rate. He headed straight towards the light at the end of the tunnel. If he ran into a barrier, he would go around, over or under it, filling more of the paper.

For twenty-five years we spent January of every year together at some University on the west coast, Santa Barbara, Santa Cruz, but mostly Caltech. The University would give us a condominium. Rose Bethe, until the last years, stayed home to look after her aging parents, and my wife Betty had to take care of our children. So I shopped and cooked, because Hans liked home cooked meals, and kept house.

In the evening before going to bed I would bring up the problems I wanted him to think about in the bath the next morning. The master bedroom always had a large bath, as we requested. Hans felt his mind to be clearest in the morning in the bath. He would come after his bath to the massive breakfast: sliced meat from the roast or joint or chicken we had had for supper the evening before, hot bread rolls, raspberry jam and lots and lots of weak tea. Hans would begin breakfast by outlining the line of attack we should use on the problem in hand. He would estimate what we could get done by noon — and he was in less than good humor if he missed his goal, because he wanted to set out for the Athenaeum by noon for lunch. During the morning he was fueled by water, lots of water, and little boxes of raisins.

Lunch was with other people, to begin with at the professors' table, when we were at Caltech, later with just me because, although hard of hearing, Hans in his later years could always understand my midwestern accent. Occasionally, when we had a visitor, whom Hans really wanted to hear, we would place him near Hans' left ear.

At home in the evening, I would cook large chunks of meat, leftovers being warmed up for a second meal or for breakfast. Generally we would have cross-checked our solutions for the problem we had worked during the day in the late afternoon, after coming home from the Institute, over tea.

Hans had done his numerical work with his slide rule, I with my $16 hand computer. Usually we agreed on the results.

In addition to the Januaries, Hans spent two springs as our boarder in Copenhagen, eating meals with us as part of family.

Usually I had a picture in my mind of where we should go; I supplied the "don't know how." Hans would consult his "disc storage" of the thousands of problems he had worked out or thought through, and then off we went!

We began with the "Equation of State in the Gravitational Collapse of Stars" in 1979[30] and we ended with "Evolution and Merging of Binaries with Compact Objects" (with Chang-Hwan Lee) which will be published in the Hans Bethe Centennial *Physics Reports*.[4]

Nobody except Rudi Peierls really worked with Hans for any length of time before I did. I saw why, because I had to wait until he was 72 before I could keep up with him. He was the leader, although I supplied the initial idea, in our wonderful paper which I shall return to below, "Evolution of Binary Compact Objects That Merge"[31] and I was frightened at the factor 20 increase in gravitational mergings that we predicted for LIGO over other predictions. Hans wrote the first draft of that paper and I had to slow him down. He was 92 when the paper was published. We carried out the common envelope evolution of a neutron star in the red giant envelope of a giant companion star analytically. I show Hans' analytic common envelope evolution in Fig. 1, on page 5424 of our collaboration. M_A is the mass of the neutron star, M_B the initial (ZAMS) mass of the giant. The calculation is explained in the original paper, also published in Bethe *et al.*[1] We needed a parameter of dynamical friction $c_d = 6$, which had been calculated by Japanese aeronautical engineers, but otherwise only Newton's and Kepler's laws were involved. Note that this is page 5424 of our joint work together.

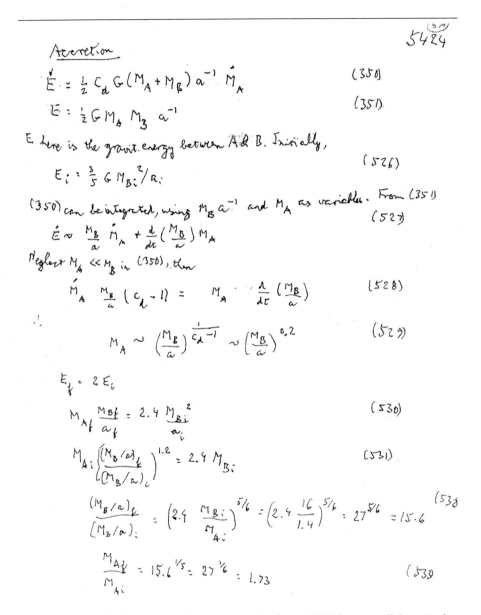

Fig. 1 Hans Bethe's handwritten manuscript (page 5424 for our collaboration).

In order to have as many equations as variables, Hans had to neglect the compact object mass M_A as compared with the giant mass M_B. This is what I meant by saying he saw "the light at the end of the tunnel." He could then solve the equations analytically. (Up to this point they had been solved only by calculations on electronic computers) and he found that the final mass of the neutron star was 1.73 times that of its initial mass, Eq. (533) in Fig. 1. In other words, for the usual neutron star mass

$$M_{\text{NS}} = 1.4 M_\odot,\qquad(17)$$

it would end up at a final mass of $2.4 M_\odot$, which Hans and I would identify as a black hole. Later Belczynski et al.[32] showed that Hans' approximation of neglecting the mass of the compact object relative to the other masses decreased the accretion by $\sim 25\%$. Their more accurate solution involved solving some differential equations numerically.

More than the fact that our calculation could be done analytically was remarkable. Until this calculation, the increase in mass of the neutron star by accretion had been neglected in calculations, although Chevalier[33] made an estimate that the neutron star would accrete about a solar mass and I[34] had not only confirmed Chevalier's estimate, but shown that there was an alternative way to evolve double neutron star binaries in which the first born neutron star did not have to go through common envelope evolution with the red giant and would therefore survive as a neutron star. However, my special way required the two giant progenitors to burn helium at the same time, which happens only when they are within 4% of each other in mass. Thus, the two neutron stars in a double neutron star binary must also be very close to each other in mass, although for the less massive giants and neutron stars this is somewhat changed by a later episode of common envelope evolution and accretion (of ~ 0.1–$0.2 M_\odot$) during helium shell burning of the remainder of the giant after it has lost its hydrogen envelope.

Research workers generally did not and still do not accept our accretion, because it exceeds by ~ 8 orders of magnitude the Eddington limit which is supposed to give the upper limit on accretion in astrophysics. Hans and I had shown in the collapse of large stars[30] that neutrinos could be trapped in hypercritical accretion, and if neutrinos can be trapped, then surely photons can be too.

Because nobody includes our order of magnitude increase in gravitational mergings from black-hole, neutron-star mergings in their estimates for LIGO, Hans and I, with Chang-Hwan Lee, prepared a manuscript, "The Evolution and Merging of Binaries with Compact Objects," for Hans' Centennial *Physics Reports*.[4] I was hoping he would be alive to take part in his centennial. I discussed this manuscript with him the morning of his death. His mind was clear and his voice was strong. (I also told him that C. N. Yang was writing a review of the "Bethe Ansatz" — Yang gave it that name "Ansatz" — for this volume. Hans was also very happy about that. He wanted his physics, not his philosophy, explained.)

As I wrote in my short autobiography "Fly with Eagles,"[35] when I was growing up in South Dakota we had a family friend, Dr. Hume, who had taken his degree in Animal Husbandry in Germany. When I was six years old, I asked Dr. Hume to his amusement, "What is the Universe?" I, never in my wildest dreams, dreamt that I would have someone of Hans Bethe's ability to work with in order to try to answer my question. Of course we didn't answer it, but we had a lot of fun and excitement in having a go at it. When Hans Bethe died March 6 I lost a father figure, collaborator and friend. The world lost its main spokesman for words rather than weapons as well as a great scientist.

References

1. H. A. Bethe, G. E. Brown and C.-H. Lee, *Formation and Evolution of Black Holes in the Galaxy, Selected Papers with Commentary*, World Scientific, Singapore, 2003.
2. H. A. Bethe, *Selected Works of Hans Bethe, with Commentary*, World Scientific, Singapore, 1996.
3. H. A. Bethe, *Zeits. Physik* **71**, 205 (1931).
4. Hans Bethe Centenary Volume, *Physics Reports*, to be published.
5. J. Bernstein, *Hans Bethe, Prophet of Energy*, Basic Books, New York, 1979.
6. A. Bethe, H. A. Bethe and Y. Terada, *Zeits. Physikalische Chemie* **112**, 250 (1924).
7. C. D. Davisson and L. H. Germer, *Phys. Rev.* **30**, 705 (1927).
8. H. A. Bethe, *Ann. Phys.* **87**, 55 (1928).
9. H. A. Bethe, *Ann. Physik* **5**, 325 (1930).
10. H. A. Bethe and E. Fermi, *Zeits. Physik* **77**, 296 (1932).
11. H. A. Bethe and A. Sommerfeld, Elektronentheorie der Metalle, in *Handbuch der Physik* **24** (2), Springer, Berlin, 1933, pp. 333ff.
12. H. A. Bethe and R. E. Peierls, *Proc. Roy. Soc. A* **148**, 146 (1935).
13. H. A. Bethe and W. Heitler, *Proc. Roy. Soc. A* **146**, 83 (1934).
14. H. A. Bethe and R. F. Bacher, *Rev. Mod. Phys.* **8**, 83 (1936).
15. H. A. Bethe, *Rev. Mod. Phys.* **9**, 69 (1937).
16. M. S. Livingston and H. A. Bethe, *Rev. Mod. Phys.* **9**, 245 (1937).
17. M. Goeppert-Mayer, *Phys. Rev.* **75**, 1969 (1949).
18. O. Haxel, J. Hans, D. Jensen and H. E. Suess, *Phys. Rev.* **75**, 1767 (1949).
19. H. A. Bethe and C. L. Critchfield, *Phys. Rev.* **54**, 248 (1938).
20. G. E. Brown, A. Heger, N. Langer, C.-H. Lee, S. Wellstein and H. A. Bethe, *New Astronomy* **6**, 457 (2001).
21. H. A. Bethe, *Phys. Rev.* **55**, 434 (1939).

22. R. Rhodes, *The Making of the Atomic Bomb*, Simon & Schuster, New York, 1986.
23. R. Rhodes, *Dark Sun: The Making of the Hydrogen Bomb*, Simon & Schuster, New York, 1995.
24. H. A. Bethe, *Annu. Rev. Astron. Astrophys.* **41**, 1 (2003).
25. S. S. Schweber, *In the Shadow of the Bomb*, Princeton University Press, Princeton, 2000.
26. H. A. Bethe, B. H. Brandow and A. G. Petschek, *Phys. Rev.* **129**, 225 (1963).
27. S. A. Moszkowski and B. L. Scott, *Ann. Phys.* **11**, 65 (1960).
28. F. Dyson, *Science* **308**, 219 (2005).
29. G. E. Brown, *Nucl. Phys. A* **762**, 2 (2005).
30. H. A. Bethe, G. E. Brown, J. Applegate and J. M. Lattimer, *Nucl. Phys. A* **324**, 487 (1979).
31. H. A. Bethe and G. E. Brown, *Astrophys. J.* **506**, 780 (1998).
32. K. Belczynski, V. Kalogera and T. Bulik, *Astrophys. J.* **572**, 407 (2002).
33. R. A. Chevalier, *Astrophys. J.* **411**, L33 (1993).
34. G. E. Brown, *Astrophys. J.* **440**, 270 (1995).
35. G. E. Brown, *Annu. Rev. Nucl. Part. Sci.* **51**, 1 (2001).

Upper left: Hans Thorner; lower right: Werner Sachs; lower left: Hans Bethe and Rudi Peierls, in 1927.

A little boy, seeing the group starting up the mountain said "Das sind Richtige Männer, die wacker auf die rauhen Berge steigen." ("Those are real men, who bravely climb the rough mountains.")

Willy Fowler, Gerry Brown and Hans Bethe, at the California Institute of Technology, 1985.

Genia Peierls, wife of Rudi Peierls, at Los Alamos, 1945.

Hans Bethe and Gerry Brown, Anyo Nuevo, 1992.

Rose Bethe dancing with Hans at the Nobel celebration, December 1967.

Hans Bethe dancing with his daughter Monica at the Nobel Celebration, December 1967.

My Life in Astrophysics*
Hans A. Bethe

Astrophysics has been an important part of my personal and scientific life three times. The first was in 1938 when I did work on stellar energy production. The second was a joyful period nearly 30 years later when that work was rewarded with the Nobel Prize in physics. And the third has lasted over the time since my retirement in 1975 during which Gerry Brown and I have had a very satisfactory collaboration exploring various aspects of supernovae and, more recently, binary pairs.

1. Introduction to Astrophysics

My first involvement with astrophysics came as a result of Carl Friedrich von Weizsäcker's suggestion to investigate the fusion of two protons to form a deuteron, namely

$$H + H \Rightarrow D + e^+ + \nu. \qquad (1.1)$$

This is obviously a beta interaction. George Gamow and Edward Teller had previously shown that beta interactions do not need to be scalar, as Fermi had originally proposed, but can also be (given) by any of four other covariant expressions. We now know that the correct expression is $\nu - \alpha$, vector minus axial vector.

Gamov suggested to one of his graduate students, Charles Critchfield, that he actually calculate the proton-proton reaction. When Critchfield had finished his calculations, in early 1938, Gamow suggested that he submit

---
*This manuscript was written with the assistance of Henry Bethe. Reprinted with permission from *Annu. Rev. Astron. Astrophys.*, **41**, 1–14 (2003). © Annual Reviews.

his paper to me because I had worked in detail on nuclei consisting of two nucleons. I found the calculations to be correct, and we wrote a joint paper.

In stars, the proton-proton reaction is usually followed by a chain of reactions with the end result of producing ^{4}He. The most common chain is

$$D + H \Rightarrow {}^3He + \gamma \qquad (1.2)$$

$$^3He + {}^4He \Rightarrow {}^7Be + \gamma \qquad (1.3)$$

$$^7Be \Rightarrow {}^7Li + e^+ + \nu \qquad (1.4)$$

$$^7Li + H \Rightarrow 2\,{}^4He. \qquad (1.5)$$

Several other chains are possible. Reactions 1.1 and 1.4 are important for observation of solar neutrinos. Reaction 1.4 is a necessary consequence of the formation of ^{7}Be. However, Reaction 1.4 starts from a preformed nucleus, ^{7}Be, whereas in Reaction 1.1 the protons have to find each other and overcome the potential barrier.

As observed above, the end result of these chains is the combination of four protons into one α-particle. This reaction releases a large amount of energy, which can be calculated from the exact atomic weights of hydrogen and ^{4}He. The rate of the reaction is determined by the first element of the chain; the following reactions are very fast.

We had a problem, however. At the time, Arthur Eddington's estimate of the central solar temperature was 40 million degrees. At that temperature, the rate of the reaction was much too high compared to the observed radiation of the sun.

2. The Washington Conference

Every spring, a small conference was held in Washington, sponsored jointly by the Carnegie Institution and George Washington University. Gamow and Teller usually suggested the subject of the conference. In 1938, they suggested energy production in stars. They invited about five astrophysicists and ten physicists, including me. I did not really want to go because at the time my main interest was quantum electrodynamics. That subject had to wait another decade to be solved.

Teller urged me to come, and I finally gave in. The conference turned out to have one really important piece of information: Strömgren, a well-known Scandinavian astrophysicists, reported that the central temperature of the sun was now estimated as 15 million degrees, not Eddington's 40. This is still the estimate. This change came as a result of assuming that the sun was predominantly hydrogen with approximately 25% helium, rather than assuming it had about the same chemical composition as the earth. The lower atomic weight of the revised mix lowered the temperature.

The lower temperature meant that the reactions calculated in the paper by Critchfield and me correctly predicted the luminosity of the sun, that is, the amount of observed radiation. So we had a theory of energy production by the sun that was immediately accepted by the conference.

This left unsolved the question of energy production in larger stars. From observations, one could show that core temperatures increase slowly with increasing mass, but luminosity increases very rapidly. The proton-proton reaction could not predict this, as the rate of the reaction increases fairly slowly as the core temperature rises.

The other key question explored at the conference was how to build elements heavier than helium. The major problem was that no nucleus of atomic weight 5 exists, nor does any of weight 8. Both immediately disintegrate, which had been shown by laboratory experiments. Elements of weight 6 or 7 in a proton sea will quickly decay into two helium atoms in reactions similar to Reaction 1.5.

Ed Salpeter, a Cornell colleague, eventually solved this problem. The principle reaction that enables crossing the atomic-weight-8 barrier is for three alpha particles to combine to form carbon. This only happens at very high temperature and density. This, in turn, occurs when the core protons are used up, and the core contracts because of gravity. The contraction is adiabatic. Both temperature and density increase until they are sufficient to start "burning" helium. The problem of how to create heavier elements is one that I have returned to in recent years.

3. The Carbon Cycle

I did not, contrary to legend, figure out the carbon cycle on the train home from Washington. I did, however, start thinking about energy production in massive stars upon my return to Ithaca. Because the observed energy production increased faster than the proton-proton reaction could explain, there had to be another reaction, and it had to involve heavier nuclei. Lithium, beryllium, and boron are the lightest elements heavier than helium. They could all be ruled out because of comparative scarcity. The next element is carbon.

Repeated reaction of carbon with protons yielded a favorable result:

$$^{12}\text{C} + \text{H} \Rightarrow {}^{13}\text{N} + \gamma \tag{3.1}$$

$$^{13}\text{N} \Rightarrow {}^{13}\text{C} + \text{e}^+ + \nu \tag{3.2}$$

$$^{13}\text{C} + \text{H} \Rightarrow {}^{14}\text{N} + \gamma \tag{3.3}$$

$$^{14}\text{N} + \text{H} \Rightarrow {}^{15}\text{O} + \gamma \tag{3.4}$$

$$^{15}\text{O} \Rightarrow {}^{15}\text{N} + \text{e}^+ + \nu \tag{3.5}$$

$$^{15}\text{N} + \text{H} \Rightarrow {}^{12}\text{C} + {}^{4}\text{He}. \tag{3.6}$$

All these nuclei are well known in the laboratory. ^{13}N and ^{15}O are positron radioactive with half-lives of approximately 10 min.

There are two basic forms of nuclear reactions in stellar energy production. One involves penetration of the potential energy barrier. The other involves a weak interaction, namely beta decay. Both of these are unlikely events, so reactions that require both, such as the basic proton-proton combination of Reaction 1.1, are significantly less likely than those that require one or the other only, which is the case in all the reactions of the CN cycle. The reactions requiring potential barrier penetration become significantly more likely as temperature and density increase; the weak interactions are essentially unaffected. The heavy nuclei involved in Reactions 3.1, 3.3, 3.4, and 3.6 have much stronger potential barriers than the protons in Reaction 1.1. The lower probability of the penetration of these barriers is compensated by the accompanying need for a weak interaction in Reaction 1.1, making the combination of all four occur with approximately the same probability as the one.

Perhaps the major surprise of the CN cycle is that Reaction 3.6 does not lead to the formation of ^{16}O in the same type of reaction as Reactions 3.1, 3.3, and 3.4. This indeed does happen, but only in about one event in 1000. The conversion to ^{16}O requires an interaction with the electromagnetic field in addition to the rearrangement of nucleons needed for either reaction. This further requirement reduces the probability by a factor of at least e^2/hc. On the other hand, Reactions 3.1, 3.3, and 3.4 do not lead to formation of an α-particle and a lighter nucleus because the binding energy of the boron and ^{11}C nuclei are low and there is not sufficient energy to permit those reactions. ^{12}C is exceptional in having a very high binding energy because it is a multiple of α-particles.

The result of the CN chain is remarkable. At the end, the starting ^{12}C nucleus is recovered, and four protons have been combined into an α-particle, just as in the chain starting with the proton-proton reaction, but through a totally different mechanism. The ^{12}C nucleus serves essentially as a nuclear catalyst, so that relatively few carbon nuclei are needed to allow frequent occurrence of the reaction.

4. Subsequent Developments

The discovery of the CN cycle took me about two weeks. I first wrote of the result to my close friend, Rudolph Peierls, who wrote back that it was a very nice result. I also told Edward Teller, who was equally pleased and congratulatory. Soon after I had finished, I gave a physics colloquium at Cornell. This excited R. C. Gibbs, the chairman of the department. He suggested to the Cornell public relations officer that I had done something noteworthy. I explained the theory to the publicity officer, and he managed to get *The New York Times* to print an article about it. Other interviews followed, and several friends asked when I was going to Hollywood to help make the movie!

More important was an invitation from John Van Vleck to speak at the Harvard physics colloquium, which I accepted happily. In the front row at the colloquium sat Henry Norris Russell of Princeton, the acknowledged leader of American astrophysics. When I had finished my presentation he asked a few searching questions. He was convinced and became my most effective propagandist.

Of course, not everyone was convinced. Some wanted experimental proof. Stanley Livingstone, who was then at Cornell, had built a small cyclotron. He ran an experiment bombarding carbon with protons. He quickly saw evidence of the formation of ^{13}N from the resulting radioactivity. Later, Willy Fowler and Charles Lauritsen of Caltech ran a more complete experiment and produced helium from continued bombardment of carbon with protons. Because of this experiment, I went down to Pasadena after teaching summer school at Berkeley. This led to a long friendship with Fowler and an association with Caltech that only ended recently when I could no longer travel.

Of course, I wrote up my discovery and submitted it to the *Physical Review*. Before it appeared, however, a very advanced graduate student Robert Marshak came to me. Marshak already had an M.A. from Columbia and now wanted a Ph.D. He already know quite a lot of physics but, more importantly for me, knew a lot about money available from prizes. For example, he knew that the New York Academy of Sciences was offering a prize for the best original paper on energy production in stars. Because the paper could not have been previously published, I withdrew my write-up of the CN cycle from an understanding *Physical Review* and submitted it to the Academy for consideration. Once the prize, $500, was safely in hand, I resubmitted the article, but this delayed publication until 1939.

The prize was very convenient. I gave $50 to Marshak as a "finder's fee." The reason he had known about all these sources of money was his own poverty at the time, and he found this fee very welcome. Another $250 was used as a "donation" to the German government to secure the release of my mother's goods because she had finally decided to emigrate.

As an aside, Marshak was interested in astrophysics. I suggested he look into white dwarfs, the final phase of the life of fairly small stars, from about one to about eight solar masses. His thesis was excellent. Years later, the Caltech white dwarf specialist told me that is still (then) served as the basis of all understanding of this type of star.

But the trip to Pasadena in 1940 was the end of my involvement in astrophysics for a long time, First, World War II intervened, and after the war, for many years I concentrated on nuclear physics.

5. The Nobel Prize

5.1. *The news and preparations*

On Tuesday, October 11, 1967, the phone rang at about six in the morning. This was about an hour and a half before my normal time to get up. It was a phone call from a Swedish journalist who told me that I had been awarded the Nobel Prize in Physics that year. The stated work for which I was being awarded the prize was the discovery of stellar energy production mechanisms. This, of course, made me very happy.

The phone never stopped ringing that morning. If it wasn't friends or family calling with congratulations, it was journalists wanting to know how I felt. My brother-in-law, who was visiting from England, was afraid that World War III had started and that it was the government calling so often. My wife was sleeping blissfully in another room. She finally woke up around 7.30 A.M. I had just enough time to tell her what had happened before the Swedish reporter Mr. Feldkirch arrived to film my day. At about the same time, a call came, which Rose answered. It was from the University wanting to know whether they could schedule interviews and a reception. Knowing my dislike of disruptions of my normal schedule and still half asleep, she responded, "Okay. The day is shot anyway." This answer made the local news, and she had to live with it for a long time.

The next two months were among the happiest of my life. It was a perpetual fair of congratulatory letters and telegrams from a great many people, including a number of Bethes who wanted to claim a relationship. Some also had plans for spending the prize money! Among the preparations for the trip to Sweden, I did something for the first — and only — time in my life: I accompanied my wife clothes shopping. She needed formal dresses for several occasions including the Prize ceremony and a dinner at the palace, and warm clothing for daytime wear while we were there: unexpected complications of winning the Prize.

The other complication was that I felt a need to catch up on the developments in astrophysics. I had not looked at the subject for nearly 30 years, and among other duties, the prizewinner is expected to talk about the prize-winning topic. Most of the time not spent in various aspects of setting up the trip was spent boning up on stellar energy. I felt like a student cramming for an exam.

5.2. *The trip*

On December 6, 1967, Rose and I were joined by our son, Henry, for the flight to Stockholm. Our daughter, Monica, who lives in Japan, would join us from there, and my stepmother, Vera, would come up by train from Germany. The Nobel experience is amazing and unforgettable. The SAS crew on the flight cosseted us. Despite a very full schedule, and attention to minute details of protocol, I never felt hurried but felt continually wrapped in comfort and a sense of my own importance. From the moment of departure to the moment of return to native soil, initiative was taken out of my hands in such a kindly way that I could just relax and enjoy.

Upon arrival in Stockholm, we were met by Mr. Rydberg of the Nobel Foundation. A very nice young man and an equally nice young lady who were assigned to make our time in Stockholm as easy as possible accompanied him. A batch of reporters also greeted us. After a fairly brief press conference — the first of many — we were whisked to the Grand Hotel in downtown Stockholm. We were given a portfolio of instructions and schedules. Both Rose and Henry were asked whether they had any special desires. Henry's request to play some bridge was arranged through the good offices of the columnist for one of Stockholm's newspapers. Rose's interest in sex education in Swedish schools was also satisfied.

The first three days were occupied with acclimating ourselves to Stockholm, renting necessary formal clothes for both Henry and me and attending various luncheons, receptions, and news conferences. Some of these conferences were fairly impromptu. We had rooms on the third floor of the hotel. My wife and I shared a suite, and the others had single rooms elsewhere on the floor. Between our rooms and Henry's was another suite. One morning Henry was awakened by a clamor on the staircase, which was directly outside our room and his. He peeked out of his room and saw about half a dozen reporters camped on the landing. Out of the middle room, a well-known actor appeared looking annoyed at the noise. He, it turned out, was in Stockholm for the local opening of his latest movie. "Oh no, sir," said one of the reporters, "we don't want you. We were hoping to see Dr. Bethe." Stockholm during Nobel week must be the only place in the world where this could happen.

The nicest of the various affairs during the run-up to the ceremony was lunch at the Nobel Foundation. Among other topics of conversation, Mr. Rydberg told me something of the selection process. Perhaps the most interesting to me was the reason that I was the first to be honored for work in astrophysics. Nobel's wife, he told me, had run off with one of the leading mathematicians and astronomers of the time. So the Prize bequest had specified that the work honored had to have practical application and that neither pure mathematics nor astronomy could be considered. Otherwise, Nobel feared, this man would have been one of the first winners. In addition, Nobel had specified that the work must not have weapons as the primary application. Until peaceful fusion power was at least a glimmering hope, stellar energy production could not be honored. Unfortunately, peaceful fusion power is still at best a very distant glimmering hope.

Another unexpected but charming event came one morning when we were awakened by three very pretty young girls clad all in white except for a crown of lit candles. They were singing a song, the only words of which that I can remember were "Santa Lucia" and offered us a choice of glögg, a warm spiced wine, or hot spiced cider. If I had had three young blonde daughters, I surely would have taught them this particular custom.

There were other planned events during the three days leading up to the ceremony. There was a luncheon at the American embassy for the three American prizewinners of the year, and a cocktail reception at the German embassy for the two Germans. The "other" German, Manfred Eigen, was an extremely nice young chemist who later became a Distinguished Visitor at Cornell. The Germans were now proud to claim me, as they had not been in 1933.

I also visited the Swedish Atomic Energy Commission. They were very proud of their plans for nuclear power plants to supply Sweden's energy needs. They stopped building them a few years later, and I wonder now how they hope to replace the power as those plants reach the end of their useful lives.

5.3. *The prize ceremony*

The prize is always given on the evening of December 10. Of course, evening in December in Stockholm is a misnomer. The sun never really gets very far up, and by 2:30 or 3 in the afternoon the sun has set, the lights are on, and it is cold and dark. So, in the early evening, perhaps around 4 P.M. or so, we were taken to the State Theater for the ceremony. This is a huge room with 1000 or more seats on the main floor and three balconies besides. The laureates and their introducers sit on the stage under the watchful eye of all the spectators and Swedish television. The families of the Nobelists and royal sit in the front rows.

Everyone is seated, then they rise as the blare of trumpets announces the King's arrival. Each winner is then announced by a short speech given in Swedish. The winner then walks down off the stage, is met by the King who hands over the medal, and then must back away to return to his seat on the stage. I found this last action by far the hardest part of the trip.

When the last of the prizes, literature, has been given, the entire crowd takes a short walk from the theater to the town hall. There, dinner is served in a gilded ballroom. The royal entourage, the prizewinners, and spouses are seated on a raised platform. Protocol puts the physicist's wife on the King's right and the writer's wife on his left. Because that year the literature laureate's wife spoke only languages in which the King was not fluent, Rose had a pleasant conversation throughout dinner. They talked mostly about Etruscan graves: Rose and I had visited some of the archeological sites, and the King had done some of the archeology. Hers was certainly the most interesting of the dinner conversations that any of us had. My own companion was an elderly princess. I hope I entertained her adequately.

After the dinner, each of the laureates had to "sing" for their supper with a short speech. I remember little about any of them except that George Wald, an American biologist, gave the funniest. He said the life of the biochemist was getting to know molecules, first small molecules, eventually bigger and more complicated ones. The most important thing is to remain friendly with the molecules.

When dinner was over, the party moved to the room next door where there was dancing until the wee hours. I even had a dance with Rose and another with my daughter.

5.3.1. *Other events*

We stayed for two more days. On Friday morning there was a television show arranged by Mr. Feldkirch with the three Americans, one of the English winners, and the Swedish winner. It was surprisingly enjoyable. In the afternoon, I gave my Nobel lecture at the Swedish Academy. In the evening, Rose and I went to a quiet dinner at the Royal Palace with the other laureates, their wives, and a few members of the royal family.

On Saturday, Rose and I went to Uppsala, home of the oldest university in Sweden. I would have liked to see Uppsala in June. In December, the trip gave us a real taste of Sweden in winter: damp, dank, and dark. Stockholm, however, had not really made us feel this way because the snow and bright street and building lights made the whole city look like a Christmas decoration. I gave my Nobel lecture again.

In the evening, the physics students at the University of Stockholm hosted us for dinner. This was a joyful affair. It was initiated into a number of Swedish customs. One that I remember is that a person must not drink wine alone. First, a man must find a woman and invite her to share the drink by looking deep into her eyes. At the time, this had to be initiated by the man. I wonder whether gender equality has extended to this custom as well. I was also initiated into the Order of the Frog. I have no idea why the frog, but I still have the ceramic frog among my Nobel souvenirs.

On Sunday, we sadly boarded SAS for the trip back to the United States.

5.4. *Speeches*

Nobel Laureates are expected to give a speech about the work that led to the Prize, first the Royal Swedish Academy, then repeated at one or more universities. I found this a fairly forbidding prospect because I had not looked at astrophysics for almost 30 years. As I studied, I decided that my lecture would only be interesting if it covered the developments in the field that resulted from understanding stellar energy production. The CN cycle is by now high-school physics!

5.4.1. Preparations

Fortunately, Ed Salpeter had kept up and was able to direct my reading during the available two months. The first thing I reviewed was nuclear reaction rates. The proton-proton reaction that is dominant in the sun increases in proportion to approximately the fifteenth power of core temperature. The CN cycle, however, increases much more rapidly in rate, in proportion to approximately the twentieth power of core temperature. The core temperature is essentially dependent on the mass of the star, so for most stars significantly larger than the sun, the CN cycle becomes the dominant method of energy production. The exception is very, very old stars that started life with no carbon and thus could only utilize the proton-proton reaction.

As long as there is a significant amount of hydrogen in the core, conversion of protons to α-particles will be essentially the only source of the star's energy. The length of time it takes to consume the hydrogen depends on the size of the star and the core temperature. The sun will last ~ 10 billion years altogether. It is ~ 5 billion years old now and has converted approximately one half its original core hydrogen to helium. Larger, hotter stars use up the hydrogen much more quickly.

When the supply of hydrogen is exhausted, there is no longer radiation pressure to maintain the size of the core, and the core begins to fall in on itself. It becomes denser and more compact. The density will increase from ~ 100 g/cc to $\sim 10^5$ g/cc. The in-fall of the core releases gravitational energy; this was proposed in the mid-nineteenth century of Helmholtz and Kelvin as the source of stellar energy. The energy flows to the surface of the core. At the surface, there is still hydrogen, which is heated to a temperature sufficient for the proton-proton reaction to occur. The energy produced at the surface needs an escape route. To accommodate this, the outer portions of the star expand and dilute; the star becomes a giant with a radius many times that of the original star. The surface temperature declines substantially because it is so much further from the energy source, and the star becomes red in the visible spectrum, a red giant.

The giant star gradually loses mass from its surface where the gravitational attraction is small. Paczynski studied this. He concluded that the giant stage continues for a long time, with the primary energy source, the CN cycle, in the shell around the core. Some of this shell is carried outward by convection and eventually escapes, thus forming clouds outside the star. Clouds have been observed near red giants that contain an excess of nitrogen relative to carbon and oxygen, a useful confirmation of the CN cycle.

In these giants, the core at the time of collapse consists primarily of α-particles, carbon, nitrogen, and oxygen. There is an abnormally high amount of nitrogen involved in the CN cycle. The collapse of the core increases the temperature. At about 100 million degrees the nitrogen begins to react with α-particles and much is converted to carbon or oxygen restoring the normal relative abundance. The core continues to contract and increase in temperature, and at about 130 million degrees three α-particles will start to combine to form carbon nuclei, thus bridging the atomic-weight-8 gap.

All stars undergo a transition to a red giant when the core hydrogen is used up. In this transition, the core contracts and the outer part of the star expands enormously. The core contraction heats the core. If the core is large enough, the temperature will get high enough for helium to combine to form carbon. The star uses up its helium also, leaving a core remnant that is almost completely carbon and oxygen. For most stars, a helium or carbon-oxygen core is the end of the road. The hydrogen surrounding these stars drifts away, leaving a "white dwarf," which is luminous because of the residual heat of the core. In larger stars, the heat generated by the contraction of the carbon-oxygen core reaches temperatures sufficient for the carbon and oxygen to combine and eventually form iron. (Interestingly, even now, with much better observation facilities, very few iron white dwarfs have been seen.)

Hertzsprung plotted the relationship between the size of stars and their absolute luminosity (Fig. 1). The long line sloping down and to the right is the so-called main sequence. The fairly flat line that breaks off about half way down is the stars burning helium. So, I learned that understanding the way in which stars produce energy had indeed allowed astrophysicists to understand the life cycle of stars and to explain many observed phenomena.

6. Returning to Astrophysics

After winning the Nobel Prize, I again abandoned astrophysics for a number of years. There was a brief excursion in the spring of 1970 when I collaborated with Baym and Pethick, both thermodynamics specialists, on a paper discussing the structure of neutron stars. But after it was done, I returned to nuclear physics.

In 1975, I retired from Cornell. At the retirement party, Gerry Brown came to me and suggested that we collaborate. Unfortunately, this had to wait a few more years, as I was deeply involved in the energy crisis at the

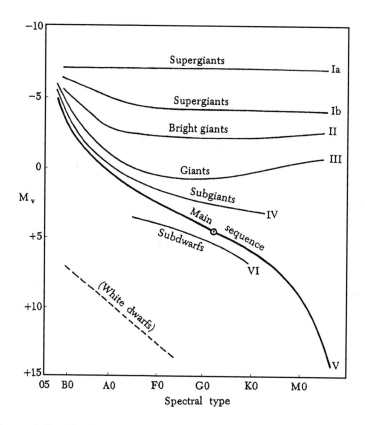

Fig. 1 Spectral classification and the Hertzsprung-Russell diagram. M_v: visible magnitudes (by astronomical definition).

time. Finally, in April 1978, both Gerry and I were in Copenhagen. "This year we will do supernovas," he said, starting an effort that would last two decades.

6.1. Phenomena

From time to time, bright objects suddenly appear in the sky, remain bright for a few months, then gradually fade away. These are called supernovae. Close supernovae are fairly rare events. Tycho Brahe and Johannes Kepler observed two around 1600 and kept careful records. The Chinese observed about ten previous supernovae, including one around A.D. 1000, another in 4 B.C. Baade and Zwicky rediscovered supernovae in the twentieth century. There is now a systematic search for supernovae, organized by the University of California at Berkeley. Approximately 50 are seen annually. It is estimated that a large galaxy like the Milky Way has about two each century.

There are two types of supernova: Type I does not show spectral lines of hydrogen; Type II does. Type II is much more common. It is the type that Gerry proposed we "do."

Type II supernovae are the funeral pyre of massive stars. In a final conflagration, the supernova radiates approximately 10^{51} erg, about equal to the lifetime output of the sun over 10 billion years. This energy release is called a foe, which stands for (10 to the) 51 ergs.

The basic phenomena of the supernova were already known. When most of the helium has been consumed to make carbon and oxygen, there is no longer any source of energy in the core. The core collapses (again). This time, heavy nucleus formation continues until the core consists almost entirely of iron. Interestingly, probably almost all matter in the universe has gone through at least one supernova. We know this because iron is the most abundant element after hydrogen and helium.

The collapse and the formation of iron produce a great deal of energy, a small amount of which escapes to produce the visible supernova.

6.2. *First efforts*

On April 1, 1978, I arrived in Copenhagen from a trip to Turkey. I had been working with increasing frustration on pions at the time. During our visit to Turkey, Rose and I had walked up Mount Pion, and I had sprained my ankle. Gerry Brown met us at the Copenhagen airport, heard this, and suggested that it was a message from the gods that it was time to change subjects. I left Rose to unpack and went to Bohr Institute. On my desk, Gerry had put the existing literature on the core collapse of massive stars. I quickly found what I thought was an error. At the time, the consensus was that the core collapse ended when the core density reached about 10% of the nuclear density. I did some calculations that convinced the two of us that the collapse continues to densities well in excess of nuclear density. Only then is the pressure repulsive and sufficient to stop the continued collapse.

We quickly wrote a paper showing the calculations and convened a hasty mini-conference in Copenhagen to discuss the result. The paper (and the conference) was a success, and Gerry and I were launched on our astrophysical career.

What we were really interested in, however, was the mechanism of the spectacular stellar explosion. Gerry and I went every year to Caltech (or some other sunny spot in California) for the month of January, and every year we would slightly revise our analysis of this bang. Every year, Gerry would return to Stony Brook and involve an associate in improving the computer model to see whether it would "explode," and every year it failed. There were several other groups working on the problem, notably Jim Wilson and associates at Lawrence Livermore and Stan Woosley at the University of California at Santa Cruz. None of them could get their models to explode either. By 1985, we thought we knew enough to write a "popular science" version of our results. It appeared in *Scientific American* in May that year.

We knew a number of things. The initial mass of stars, known as **zero-age main** sequence (ZAMS), that become supernovae is at least eight times the mass of the sun ($M_\odot$). We knew that smaller stars that become supernovae, up to about ZAMS = 15 $M_\odot$, leave a neutron star as a sort of tombstone. We knew that the maximum mass of a neutron star is ~ 1.5 $M_\odot$.

It is possible to calculate the amount of gravitational energy released in the collapse to a neutron star. It is ~ 400 foe, $\sim 10^{54}$ ergs, a prodigious amount of energy. The nuclear binding energy is ~ 20 foe. The visible released energy is ~ 1 foe. The balance of the energy is stored as thermal energy of electrons and neutrinos trapped in the neutron star. Approximately 300 foe escape in the form of neutrinos during the next minute.

6.3. *1987A*

In January of 1987, Gerry and I decided to abandon the supernova explosion. We felt that there was no more to do without some observations. The last observable supernova explosion had been almost 400 years earlier. He and I began some other, nonastrophysical work. As we parted, Gerry said it was time for an observable, nearby supernova. As if responding to our request, on February 23, 1987, one appeared in the Southern Hemisphere sky a mere 150,000 light years away. This supernova became known as 1987A. One of my great regrets is that I could no longer go in person to look at it.

In any case, it came at just the right time. Like us, almost all the groups looking into supernovae had reached a dead end. However, the tools for good observation were only recently in place. Two neutrino observatories, one in the Pittsburgh Plate and Glass Mine in Ohio and Kamio-kande in Japan, both reported seeing extra neutrinos at just the right time. All sorts of telescopes, any that could be, were immediately trained on the supernova. Reports of the observations came quickly. The observations both confirmed much of what we had surmised and contained surprises.

The greatest difficulty for theories has been to convert the rapid collapse of the core of the original star into the rapid expansion of the outer material in a supernova. By 1990, many of us who were working on the problem had concluded that convection plays a major role in the process. The observations of 1987A seem to support this idea. The best formulation of this, in my opinion, was by a young physicist, Herant. With Benz and Colgate, he showed that convection could actually transport energy from the surface of the core outward in approximately the necessary amount. He did not, as far as I know, tackle the problem of getting the energy from the core to its surface. Supernova theorists produced many papers to explain what had been seen. Jim Wilson and I proposed that neutrinos and radiation carry energy from the core to the surface.

Notwithstanding the further insights into the mechanisms of the supernova gained after 1987, as far as I know no one has built a model that "explodes." So there is still work to be done.

Another aspect of the supernova that still needs to be explained is the so-called rapid (r)-process. Supernovae are the only possible source of elements heavier than lead. We find them in nature, so we know they must be made. The slow (s)-process explained by Margaret Burbidge, Geoffrey Burbidge, Willy Fowler, and Fred Hoyle explains most of the creation of the majority of the elements between iron and lead. Some of the known isotopes of these elements could not have been produced by the s-process and must come from the r-process. I have explored this from time to time, occasionally with great hope that I have had an insight, but each time have proved myself wrong.

In any case, Gerry and I worked for about nine more years on supernova theory, producing various interesting results, but never a model that exploded.

6.4. *Gravity waves and binary pairs*

By 1995, after 17 years, we had again exhausted our capacity for new insights into the supernova. During our annual visit to Caltech in 1996, Kip Thorne asked us for help. He was, and is, working on LIGO (Laser Interferometer Gravitational-wave Observatory), an apparatus designed to detect the gravity waves predicted by relativity theory. He wanted to know how often he might get a detectable event. The most promising source is the merger of similar-size massive objects. Several people had developed estimates for the frequency of the merger of two neutron stars. Kip asked us to look at mergers of small black holes with neutron stars.

"You guys," he said, "are very good at calculations of things that nobody has ever seen." This was a nice problem. Within a fairly short time, we were able to show that such pairs are much more common than neutron star pairs. By the time Gerry and I were finished with the paper, we were able to promise Kip that, instead of a couple of waves a year, he would be able to detect a few per month.

This work led to a more general interest in binary stars. Over the next few years, Gerry, with some assistant from me, looked at various types of pairs. We looked at large black hole-neutron star pairs. We looked at white dwarf pairings with neutron stars. We looked at X-ray and γ-ray burst black hole systems. Most recently, we have been exploring the effects in binary systems of a small companion within the envelope of a massive giant star, i.e. common envelope evolution.

7. Conclusion

I have now spent over a third of my working life, the past 24 years plus brief periods before that, working in astrophysics. It has been a true joy trying to understand and explain the rarely or never-seen phenomena of the universe. I thank friends and colleagues who shared insights, results, inspiration, and mental perspiration. I especially thank Gerry Brown for his unwavering support. Without him, I would never have begun, or continued, the scientific journeys I have taken in this quarter century.

Three Weeks with Hans Bethe
Christoph Adami

Hans Bethe started visiting the Kellogg Radiation Laboratory, accompanied by Gerry Brown, every winter starting in the late eighties. My first visit to Kellogg was in 1991 as a final year graduate student. I returned in January of 1992, while a postdoc with Gerry Brown at SUNY Stony Brook, and this would be my first of many winter months spent in the company of both Hans and Gerry. The following notes are literally "mental downloads" that I wrote down every evening after leaving dinner with Hans, Gerry, and sometimes other invited guests that year. I indiscriminately wrote down everything that I could remember from that day, whether it seemed important or not, hoping to edit it later. In the end, I decided to keep the editing very minimal in order to preserve the original text's immediacy and spontaneity. Besides grammar and punctuation, a few facts that I remembered incorrectly were restored. I also added references and footnotes that add explanations where necessary. The following is a transcript of the conversations that took place between 20 January and 20 February, 1992.

First Week

Monday, January 20th

I meet him for the first time in the afternoon. Gerry comes up to my office to tell me that there will be a discussion session with Grant Matthews from Los Alamos on problems of the early universe. I know that by that time, Hans has returned from Santa Barbara, and will be waiting in his and Gerry's joint office on the first floor of the Kellogg Radiation Laboratory.

I am, while walking down the stairs, slightly nervous, because of all the anticipation of meeting him. When I enter the office with Gerry there is only him, Matthews hasn't arrived yet. Gerry introduces me to Hans, who sits on a chair. His handshake is weak. I mumble that it's a pleasure to meet him. What impression do I have? He is old, but I knew what he would look like. His hands aren't trembling. He only has a little hair left around the back of his head, white, and almost no eyebrows. Yet his face, while wrinkled, is still somehow strong, probably because he's a good eater. When he leaves to get himself a coffee, he walks upright. He munches on raisins. Gerry notes that Matthews hasn't arrived and says that he'll come and get me once he does. I go back upstairs to my office. Five minutes later Gerry is back, to tell me that everything is set. Grant Matthews talks about inhomogenous nucleosynthesis. Hans listens attentively. Somehow, I can't really believe that Hans understands all this, him being 86. And every time he starts to make a comment, or ask a question, I expect to see that he cannot follow. However, that never happens. When he talks, he talks slowly, but surely. After maybe an hour of discussion this feeling slowly fades away. Hans knows what he is talking about. Nevertheless, he is very tired from the drive from Santa Barbara to Pasadena (he drove a rental car, all by himself), and around five they leave for the apartment. I join them at 6:30 for dinner. When I come in, Gerry tells me that dinner will take a bit longer (he is making a roast beef) and I sit down, across from Hans. Almost immediately, Hans addresses me:

"I have a request!," he says. I look very subservient, not knowing what will follow, and answer — "Yes?"

"Please," he continues, "can you speak at half the speed, and twice the volume? I am hard of hearing and I have this hearing aid." I am slightly embarrassed, as during the discussion in the afternoon I talked, as I always do discussing physics, admittedly very fast. I even remember worrying about it slightly, but then got swept up in the subject again. I apologize for that, Gerry makes the usual remarks about my fast-talking, and I try to speak slowly, and loudly. The conversation is not easy, I don't dare to ask questions, and so I only answer his, about where in Germany I'm from. I tell him about Brussels (where I was born) and about Bonn University, where I studied physics. He asks me about who is still on the Bonn faculty. I start with mentioning Rollnik, my advisor, but that name doesn't seem to ring a bell. I mention that he's on many committees, and now heads the one responsible for restructuring the East German physics departments. We discuss this a bit, then I mention Konrad Bleuler, knowing full well that he must remember him. They must be roughly the same age. Indeed he does remember, his face lighting up slightly when I mention him. Then I mention more names from the Nuclear Theory group in Bonn, but none are familiar to him. Then he asks me if Wolfgang Paul is still there. Yes, of course, I answer, and tell him the anecdote about how they relegated him to a little office close to the janitor's closet in the cellar after his retirement, only to reverse this move after he won the Nobel prize. Hans laughs, and goes on to tell me that he first met Wolfgang Paul when he visited Werner Heisenberg in Göttingen, in 1948. I go on to tell the story of how Wolfgang Paul received the money from the German government to build the 500 MeV electron synchrotron in Bonn, the first of its kind in Europe. Namely, he simply asked Heisenberg, who was then in charge of distributing research funds in Germany, for the money personally, and a week later it arrived on Wolfgang Paul's private account, just like that. Hans is amused, and says what a fine experimenter, and a nice person, Wolfgang Paul is. We all agree. I now remember another story about Wolfgang Paul, just as an aside, that I learned during my time in Bonn. Indeed, Wolfgang Pauli visited the institute, and under some circumstance Paul whispers to Pauli: "Finally I meet my imaginary part!"

We talk about more things, discuss the reunification of Germany, and he asks me how the East is doing economically. I talk about unemployment, salaries, etc. He seems genuinely pleased to hear that I expect that the Eastern part is slowly recovering. I tell him that the German government poured over a hundred billion marks into the East, which leaves him very astonished. Yet, the conversation is still difficult, as I mostly answer his questions. There are periods of silence in between, where I don't quite know where to look. Finally, dinner is ready. During dinner we don't talk much. Hans eats slowly, but he eats a lot. After dinner, Hans retires as he is tired from travelling. Gerry walks a piece of the way back with me as he wants to do some grocery shopping, and says that Hans and I still have to get to know each other; that today the discussion was somewhat superficial, but that he likes to talk about Germany.

Tuesday, January 21st

At around 11 a.m., Gerry calls me for discussion, now with George Fuller from San Diego; later Grant Matthews joins in. Hans is a lot livelier. I'm observing him all the time. He gets up to write on the blackboard, he interrupts George Fuller: "I love the r-process, really, but I think we don't have to worry about this here." Fuller is slightly embarrassed, he won't mention the r-process again. This session is about neutrino physics, specifically neutrino oscillation and transformation. I am not an expert on this, but it is very interesting. I learn that Hans has in his mind a pretty well-rounded picture of what the masses of the three neutrino species should be. Considering that the rest of the world is still wondering, or assuming that they probably all vanish, I find this quite remarkable. For the record: he estimates the electron neutrino to be basically massless, the muon neutrino to have a couple of milli electron volt, and the tau neutrino to have between 35 and 70 electron volt. Fuller and Hans agree that this should be enough to close the universe. Hans says: "I don't know much about inflation, but Alan Guth tells me that you need $\Omega = 1$, and here we give him $\Omega = 1$; he should be satisfied!" The subject turns to experiments to detect the μ–τ neutrino oscillations, and Hans gets emphatic: "The only experiments they do are at Fermilab, but they are doing the wrong experiment! The conversion length is a few meters" (he and Fuller just estimated it on the board with Hans's formula), "but they are looking for kilometers! They will never find it that way!" He sits down again, and calms down. "I will have to write them a letter." With that he goes to lunch with Fuller and Matthews. Gerry and I go to the Kellogg "Journal Club," to listen to a talk about solar neutrino detection.

Later in the afternoon, Gerry calls me again, for more discussion with Fuller. This time, it is nucleosynthesis again. We are having a very lively discussion, but when I speak I'm always concerned about whether I'm not too fast for Hans. I try to talk slowly — have to slow down many times — but it's very hard to control. We talk about the QCD phase transition, and I tell them that I still suspect a first-order transition involving the electric gluon condensate. Fuller is very interested, Hans is a bit skeptical, but asks the right question: "At what temperature?" I answer that I suspect at the chiral restoration temperature, Gerry objects, I concede that it's speculation, and Hans nods. I try to give an argument why chiral restoration temperature is possible, but Hans interjects that we won't solve that problem today. He wants to go home. Dinner is at 6:30.

I arrive on time. Hans already sits at the table, with food on his plate. He makes no compromises in matters of food. During dinner we try to find the Italian word for "vegetables," as Gerry cooked chicken with vegetables in an Italian way, and tries to say something like "polio al vegetabile." Hans suggests "verdura," but concedes that that might mean "salads." He goes on to say that he can guess the Italian equivalent of English words most of the time, and tells the story where, when he was in Italy, he wanted to buy some laxative, and asked for "purgatorio." The pharmacist smiled, but understood, and told him it's "purga." We talk about southern Italy, and how Pierre[a] always maintains that he is tall for an Italian, yeah, maybe with respect to those in the South! Hans mentions that many Italians that he has known were quite tall. Fermi, he concedes, was considerably shorter than he, adding that at the time though he (Hans) was several inches taller than he is now, and that Fermi had an assistant (Rasetti) who was taller than he was then. We also discuss a little bit of physics, specifically Hans wants to hear about the talk Gerry and I went to that afternoon, about solar neutrino detection. I tell him that the girl who gave the talk, a first year graduate student, reported that they see too few neutrinos, just like in the chlorine experiment, but that these are low energy neutrinos from the pp fusion reaction in the sun. Hans of course obtained the Nobel prize for his theory of energy production in the sun, which predicts many more neutrinos then observed on the ground. Gerry says that Hans is pleased with this result (too few neutrinos) because of his new theory of neutrino transformation and oscillation that predicts just that, and goes on to joke that Hans always tries to grab at any straw that would prove that he didn't get the Nobel prize for a wrong theory. Furthermore he says that the SNO experiment will really decide about that, and Hans agrees.

After dinner, Gerry goes to lie down on the couch to have a short nap, while Hans is still finishing his chocolate cake and chocolate ice cream dessert. It is his third helping. I ask Hans about different solar neutrino experiments, and eventually we get to talk about the SNO experiment, in northern Canada. He explains the process, the technology, the difficulties, the one thousand tons of heavy water that are needed for that, at $200 per liter of heavy water (loaned from the Canadian government) and marvels at the neutrino detection efficiency. We discuss its applicability to supernova neutrino detection, and again he is enthusiastic.

Somehow, the topic returns to Germany. Hans says he grew up in Frankfurt, and still likes that city. He asks if my parents still go to Germany, and I answer: of course, since many of our relatives still live there. I tell him that my father's family lives in the "Pfalz" (Palatinate), and he mentions that he also has some relatives there. I also mention that my mother's family lives in Ludwigshafen, and talk about where my parents come from originally. Then, as I realize that he likes to hear the names of German towns, I tell him that my father's family lives close to Germersheim. At that moment, his whole face lights up, and he exclaims:

---

[a]Pierre Pizzochero, Gerry's former graduate student at Stony Brook.

"Germersheim! One of my earliest childhood memories is associated with Germersheim." After a slight pause he goes on: "I was born in Strassburg, and when I was six, my father obtained a call to a professorship at the University of Kiel. So, we took the train to Kiel. It was a night train, with sleeping compartments, and I was very excited. In the middle of the night, I looked out of the window, and saw the train station sign announcing Germersheim go by. This is all I remember, I don't know why this has remained. I was very excited, a little boy travelling in a sleeping compartment."

"When I was eight," he continues, "I moved to Frankfurt." I ask him whether he had ever returned to Frankfurt, after the war.

"Of course," he smiles, "the planes land there. Also, I returned after the war to visit my father, who still lived there. Also when I visit my sister in Neuwied, it's easiest to go by Frankfurt."

We talk a bit more, in a fairly relaxed way, about Brussels. He mentions that he went there a couple of times; for example for the 1961 Solvay conference, and we both marvel about the Grand' Place. He tells me how he always thought it unreal that people actually lived in the houses around the Grand' Place (Brussels' central square), and I'm surprised, as this is what always goes through my mind when I walk over there. I tell him that, and we seem to have a genuine understanding. I relay to him similar impressions that I got from Regensburg in Germany, from little streets with the facades of houses that seem to lean inwards. I tell of cobblestones, and of the marks that the axles of horse-drawn carriages have left at the corners of the streets, hundreds of years ago, and that, on a silent night, one can mistake this century for another, earlier one. He listens to this in genuine delight. We talk about the history of Regensburg, and how it became to be a "Freie Reichsstadt." I add that its decline later was probably due to its losing this status, and he interjects that this was probably due to Napoleon. We talk a bit more about Germany, and a bit of physics. I see that he's somewhat tired, and say that I should go back to work. On the way out I see that he picks up the more interesting parts of the newspapers lying around on the couch table. Gerry told me earlier that Hans likes to read sometimes several hours a day. Before going out of the door I raise my hand and say loudly: "See you tomorrow." He smiles and waves his hand.

Wednesday, January 22nd

A quiet day. We meet at 11 a.m. for discussions. This time around we have Edward Shuryak from Stony Brook. It's an interesting discussion but nothing really stands out. Hans seems somewhat unconvinced about what

Shuryak presents, I on the contrary find it quite impressive. I could always relate to Shuryak's way of thinking. Lunch at the Athenaeum: Hans, Gerry, Edward, and I. We all have "Blanquette de veau," which is on today's menu. I note that it's a Belgian speciality. While not being absolutely sure, I assume that no one is going to contest it. From then on the subject is language, namely Dutch first, whether I speak it, etc. Hans asks me where I went to school in Brussels, and I tell him about the European School system.

Not much discussion during lunch. Gerry wants to show off Hans in front of Edward by telling some anecdotes, for instance the one about this problem in radiation from electrons, when Weisskopf got interested. Bethe told Weisskopf then: "Sure you can solve the problem, it will take you three weeks. I can do it in three days." And so he did; it is the Bethe-Heitler formula now. Then we come to speak briefly about Hans's time in Italy, and Hans mentions a paper he wrote in 1931 with Fermi, on the quantum mechanical interaction of electrons. Gerry then mentions another famous work of Hans's, solving, I believe, the problem of the magnetic spin chain, with the now so famous "Bethe ansatz." Gerry smirks that Fermi was totally uninterested in this work, and asks Hans if Fermi ever read that paper (as they were collaborating at that time). Hans shakes his head: "I believe not." Edward interjects that he finds that strange, seeing that one of the last papers of Fermi was the famous Fermi-Pasta-Ulam problem, from 1952, involving chains of nonlinear harmonic oscillators. They tried to solve that numerically, but couldn't get the system to equilibrate. This, it later turned out, was due to them having a set of parameters which made the model close to exactly solvable, of the Toda type. This of course was realized much later. Anecdote: at the end of Hans's paper on the one-dimensional spin chain he says that the solution to the two- and three-dimensional problems are "forthcoming." Needless to say, they still are. Hans asks if anybody ever solved them, and Gerry answers negatively. Hans shakes his head: "After all this time!" After lunch we grab our coffee mugs and have more discussions. Gerry mentions on the way to the office that the Soviet (Russian!) Embassy had called to tell Hans that they had a letter for him from President Yeltsin. I make a mental note to inquire about that at dinner.

It is 6:45 p.m., and Gerry has made pork loin roast. He gets to be pretty good at this cooking thing. We don't talk about physics at all that evening, (at least not with Hans). The subject is college, generated by Gerry having been called by his wife Betty on some affairs involving their son Titus's college plans. Specifically, we discuss Stony Brook as an option, and when Gerry says that it is not that easy anymore to get into Stony Brook, that they had to turn hundreds away last year, I just mention that probably they raised their academic requirements such that they don't admit anybody with a criminal record. Gerry answers dryly: "Sure they admit criminals." Then I go on to relay my first-year teaching experience at Stony Brook doing PHY 100, and about the person having all the trouble in the world to obtain the result for one thousand divided by one thousand. Hans hardly believes it; during the story I constantly have to make it clear that I'm not making it up. Hans is amused. Somehow we move to the subject of Hans's family again, in fact, it is a consequence of Gerry mentioning that he had a half-brother born in 1899. Hans mentions that he also has a half-brother, who is 27 years younger. He tells about his father having a "new batch" at age 61, says that it was nice when the children were small, but that it was hard when they were teenagers, as their father then was "decrepit," as he puts it. He also talks about the hardships of war during that time, his father's house in Frankfurt having been bombed such that they had to move to a small town a hundred kilometers north. Also, he says, the Nazis didn't like his father ("That feeling was mutual," he adds) and forced him to retire early. After the war, however, the Americans gave him back his job as professor of physiology at the University of Kiel, at age 74. He says that there were a lot of students, but not much material or labs. His father's only help was the lab assistant. Gerry throws in the anecdote of the assistant of Sommerfeld, in fact the machine shop assistant, who owned a ski lodge in the mountains, where Sommerfeld would always take his guests.

Finally, I ask about the letter from Yeltsin. Gerry had told me earlier that evidently Hans had written a letter to Boris Yeltsin with his idea of how to make the Ruble convertible. So I ask what Yeltsin's reaction to his proposal was, and Gerry jumps in and asks, "What proposal was that anyway?" So Hans begins:

"I guess now I can talk about it. I was thinking about why the Russian economy wasn't working, and I reached the conclusion that nobody has any confidence in the currency. If somebody is producing goods, then he's not selling them because he doesn't trust the money he obtains. So I thought about the experience we had in Germany in 1923, when everybody had lost confidence in the Reichsmark, and the value of the money changed by twelve orders of magnitude. Then they introduced the Rentenmark, and backed it by the value of the agricultural land in the country. While success was not expected, the scheme actually worked, and the $1:10^{12}$ ratio remained approximately constant. So I wrote to Yeltsin that they should do the same, but not back it this time with land, but with their oil, as they have huge oil reserves." After discussing this ploy a little bit (it turns out Hans knows a lot about economy and finances; one of his sons, Henry Bethe, is a vice president with Chase Manhattan bank, and the inventor of the "variable return preferred stocks," an instrument that it seems was of some importance some time ago). Gerry recalls how astonished Edward Shuryak was when Hans was picking up the telephone and it transpired that he was speaking with the Russian Embassy, answering them that, yes, he would like a translation of Yeltsin's letter as he couldn't read Russian. Gerry maintains that Hans did this just for his ego, that he timed it perfectly to show off to Shuryak, and Hans grins.

We talk a bit more about money management (Gerry likes to talk about it) and pension funds. I throw in my two cents' worth, and Hans is amazed. "Where do you know all this from?" "Well," I answer, "I just started this job as a postdoc, and they gave me a two hour lecture about TIAA/CREF and the like, but it's not as if I am thinking about retirement." "Yes," he says, and grins.

After dinner Hans says good night and disappears in his room with a book. I have another discussion with Gerry about Shuryak's talk, then go back to the Lab where I'm sitting right now, writing this.

Thursday, January 23rd

No visitors today. At noon we have lunch at the Athenaeum: Hans, Gerry and I. I don't remember much from the discussion now. We talk about recent news, Algeria for instance, and about the difficulty of choosing sides. Hans has his plate heaped up with what looked to me like legs of some animal, while Gerry and I have fish that seemed miniscule in comparison. Hans calls his dish: "Quite a responsibility." After having gossiped about Ismail's[b] recent marriage, Gerry throws in that Rose Bethe's mother was, as he sees it, very much in favor of her daughter's suitor, and Hans agrees, revealing that he had already known Rose when she was a twelve-year old girl, and was often asked to take her for a walk. Also he quips that Rose often would tease him by saying that he would really have wanted to marry her mother. So the subject changes slightly to Hans's wife's mother, Ella Ewald, who in turn was the wife of the famous crystallographer Paul Ewald. They talk about Ella as if she were still alive, so I ask, and indeed she is. Gerry mentions that she had her 100th birthday last year, and that many of her sons and daughters came, all between 70 and 80.

---

[b]Ismail Zahed, professor of theoretical physics at SUNY, Stony Brook, and my graduate thesis advisor.

Gerry somehow changes the subject to mention how he became famous, when a certain "Professor Sucher"[c] published a paper on the autoionization of atoms, mentioning in a footnote that the effect and its calculation were already known, and done by Brown and Ravenhall in 1951.

"Well," I said, "this happened to me too, but the other way around." Not without trying to impress Hans, I tell about my work on charmonium suppression in havy-ion collisions, and how it is the exact analogue of an effect described in the book "Quantum Mechanics of One- and Two-Electron Atoms," by Bethe and Salpeter (which I don't mention as it is obvious), that book having been written about forty years earlier, I presume. But I tell Hans that I just had to change the differential equations in his book slightly, the rest was almost the same. I tell him that the result was that the charmonium states get ripped apart at the instant the colorelectric fields are of the same order of magnitude as the QCD string tension, and he quietly responds that, yes, this had to be expected, and Gerry agrees. I mention that this work proceeded very quickly, in about a month, and Gerry counters that I'm impressing nobody with one month, as Hans calculated the famous 1020 megacycles of the Lamb shift in the train to Schenectady in New York coming back from the Shelter Island conference (I believe this was in 1947). Gerry mentions the names of everybody who was at that conference (all the big shots) and how they discussed this new thing "Quantum Electrodynamics," and at the end of the conference agreed that one should calculate something. Gerry goes on to say that all the big shots weren't particularly pleased that Hans got the jackpot immediately after the conference, on the train no less, but Hans replies that this is probably not quite accurate, as many people expressed the feeling to him that they were very grateful, and that without this result the others might not have pursued the matter. I have to mention that this calculation of the Lamb shift is one of the milestones of quantum field theory, and to accomplish this calculation without the technique of Feynman diagrams borders on the impossible. Yet it was done on the train, and, of course, correctly.

Dinner was scheduled at 6:45 and was ready at 7. We start by discussing the politics of the election year, and Democratic candidate Bill Clinton since he appears on the cover of *Time* magazine (Headline: "Is Bill Clinton For Real?"). Hans postulates that the Democrats can only win with a candidate from the South: "They have to carry the South." We talk briefly about Mario Cuomo whom I favor for the 1996 elections, but Gerry gives him a bad record in funding education, seeing that the State Universities have to

---

[c]Joe Sucher, professor of physics at the University of Maryland.

take a 15% cut this year. Hans and Gerry blame it on him, having reduced taxes last year, and having spent too much in the prosperity years. I disagree somewhat, as it makes no sense to not spend money when it's there, to not repair this or that bridge because there could be worse times. In bad times, one borrows. We're drifting into economy again, via a discussion of interest rates. Gerry maintains that after the recession the US will tumble into another inflationary period. "Mark my words," he says. He and Hans agree that Germany is very careful with its interest rate policy, and I remark that the country with the most stable currency in the EC (European Community) nowadays is Spain, owing to its very high interest rate. Hans is surprised, and Gerry nods, saying that in general Spain is doing very well, that it reminds him somewhat of the US in that businesses are sprouting very quickly, and only the fittest ones survive. We talk about the influence of Franco, and Hans says that he is very grateful to Franco for some things. I wonder what this could be and ask him, and he says it is a little known fact that Franco stood up to Hitler in the Second World War. He continues to tell the story (Hans is clearly a history buff) of how Hitler proposed to Franco that he should let him march unhindered through Spain, and that he would conquer Gibraltar for him. Franco refused, very wisely we all agree. This is all the more astonishing, I remark, as Hitler helped Franco in the civil war, bombing Guernica with his Air Force in 1937.

Somehow, Gerry brought the subject onto Wolfgang Pauli; I don't remember what was in between. In his typical manner Gerry remarks that: "Pauli was a smart guy." I am about to make my usual remark that he might have been smart but that he didn't publish much original stuff, when I realize that Hans probably knew Pauli personally, and thus would have a *slightly* better judgment of this fact than I, so instead I say: "I suppose he was," and turn to Hans. He, however, doesn't say anything. Then Gerry continues that after Pauli had had Weisskopf as a postdoc for one year, he was walking around shaking his head saying: "If only I had taken Bethe!" So I ask Hans if there was a choice and he says no, he never offered him the job. Gerry asks if he would have taken it, had it been offered to him. Hans thinks for two seconds then says: "I think not. I would not have survived it." We don't say anything for a while, then Gerry changes the subject. We didn't ask him why that would be so.

During dessert the word "supergravity" falls, and Hans asks me spontaneously what it is. So I try to explain to him what little I know about supersymmetry, gravity, and its unification, being slightly aware that I probably should not venture too far into gravity. He is satisfied at one moment, but I am not, as I remember that there was a bit more to supergravity than I could just summon up from my memory. However, I do tell him the story (that Peter van Nieuwenhuizen had related to us students at Stony Brook) of how they[d] discovered the supersymmetry of their theory at three in the morning, when they were all huddled together staring at the result of their computer-generated variation, gasping at every zero after zero appearing on the computer output. Hans likes that story, since it has this unmistakable scent of discovery to it. Later, still during dessert (remember Hans eats a lot) I tell him about the new CERN collider project, the LHC, and its competition with the SSC. I suppose we arrived there from my mentioning that supersymmetry might be found at these next-generation colliders.

While Gerry talks to Betty on the phone, Hans asks me whether I would prefer a (postdoc) position in Europe or in the States. I answer that it is basically a toss-up, and that I applied (as serious places) only to CERN and to Caltech, and would like to have the opportunity to choose. He agrees that it is always better to have a choice.

Friday, January 24th

I am writing slightly later today, as tonight was "story-telling night," so to speak. Nothing to report during lunch, as we had the scheduled Friday lunch of the whole Kellogg Radiation Lab group at the Athenaeum. Hans skips those because he doesn't like crowds, as he can't hear if too many people are speaking at the same time. I went to work out in the afternoon, and arrive at Gerry's and Hans's apartment at 6:30. Hans is on the phone with his wife, and when we sit down to eat at 7, (Gerry made three big trouts) he tells us that he just heard he has a new grand nephew. For part of the meal we talk about Hans's grandchildren, and we also talk (or better, I tell) about the way history was taught at my school, this on the tail end of a thread that began with why I didn't have to go through military service in Germany. Hans asks whether I felt any animosity from the Belgians because of my nationality (a question that followed a discussion of the German habit of marching through Belgium, and German war atrocities in Belgium), which I answered negatively. He also wonders whether history was taught to us in a biased way or not. Hans is delighted about my description of history class at the European School in Brussels, namely learning facts and thoughts through studying contemporary documents, and putting

---

[d]D. Z. Freedman, P. van Nieuwenhuizen and S. Ferrara, *Phys. Rev. D* **13**, 3214 (1976).

them in perspective. He said that this was very different at the time when he went to school, when it was all just dates of wars and battles.

He asks how my father went to school, so I tell him about his first years in school in Marienfeld in Romania, about how they were forced to learn the Romanian language, to when they had to flee from the Russians in 1944. I tell him about the vagaries of that time just as my father had recounted them to me, of grenades, pistols and rifles lying alongside deserted streets, with the kids picking them up to play; of carriages that smelled atrociously, with bodies falling out at times and nobody stopping to pick them up. While I'm at it I tell the story of my mother's flight from Silesia (now in Poland), of the house they left, thinking they would be back soon, the dowry for the girls that her mother had been sewing and knitting for years, all in the big suitcases that they had brought to the train station, only to have them stolen by a man who pretended to help with the luggage. Hans and Gerry are listening quietly. I tell of the children wearing whatever they could put on, all clothes on top of each other in layers, and now, with the mother crying helplessly, the father imprisoned in a war camp, the oldest daughter, my mother at age ten, having to take charge and ordering everybody into the train. It was to be the last train out of the city, the night train, and the war planes were circling like vultures. With every light out, the train thunders through the night, westwards at full speed, under enemy attack, glass shattering under the airplane's strafing fire, and children wailing crouched under the seats and tables in narrow compartments. Only the fire gushing from the overheated smoke stack betrays the fleeing train, visible to the frightened children as the train races through bends on the track. I finish the story and there is a little moment of silence. Then Hans says that, of course, they would use the planes to intercept the trains, as it was the only remaining means of transportation, the trucks having run out of gasoline. And trains were easy targets too.

"Of course," he begins, "I saw these things from the other side. I was very lucky in a way." Gerry interrupts to ask Hans to tell the story of how it all started when he was on vacation in Switzerland, in 1933. Somewhat reluctantly he begins to speak, says: "not all the details," but goes on:

"I was in Switzerland, and I get a postcard from my graduate student, it might still be preserved, and it reads: 'You have just been fired, what am I going to do?!' " Gerry asks what kind of position he held then, at the University of Tübingen, and he says that he was a lecturer giving classes in electrodynamics and nuclear physics. Later Hans tells the story of how before he got fired he was supposed to give a talk on the new paper by Chadwick on the discovery of the neutron, and how after some Nazi students were threatening to disrupt his lecture, Gerlach and another physicist suggested he should abstain, and he did. Hans continues how on his return from Switzerland they gave him the formal dismissal letter, mentioning that he would still get this month's (April) salary, but that he should pack his bags and go to his mother. He then went on to Munich where he obtained a position as Privatdozent. However, he was dismissed from this position soon after. I take the opportunity to ask Hans if they ever gave any reason for those dismissals.

"Oh," he said, "my mother was Jewish; this was all the reason they needed." I express my feelings how incredible this all appears to me, and he picks that up:

"It sounded incredible to my father too. He sat down and wrote a letter to Sommerfeld, somehow this letter is preserved — Sommerfeld kept it — and asked for advice. Sommerfeld was a very nice man, and helped all of his former students, so he wrote to Bragg at the University of Manchester, and Bragg offered me a one-year position there."

"There," Gerry interjects, "you learned how to live, in the care of Mrs. Peierls. She showed you how to cook, to wash your clothes, to clean the pans ..."

"Yes," Hans smiles, "and they were some of the most productive years, these years 1933–1934. It was such a great atmosphere, so much going on! So very different from the time in Tübingen and even Munich." Gerry mentions some of his works from that time:

"You did the photodisintegration of the deuteron, the electron problem,"

"Bethe-Heitler!" Hans throws in and literally beams.

"The theory of alloys!" Gerry continues.

"That was an idea of Chadwick," Hans interrupts again, as they exchange a battery of works and ideas.

"I never met Chadwick," Gerry reflects, "but he seems to have been a nice man." Hans objects:

"He was a sour man. He was nice as a person, but he was sour, a queer personality. Very unlike Bragg, who was charming." I ask whether Chadwick also worked in Manchester, and Hans answers no, that he was at Cambridge.

"Peierls and I would take the train to Cambridge, it was a terrible journey, the trains were so slow. Once, Chadwick had invited us and told us that they were doing these new experiments, and they were getting results, but he bet with Peierls and me that we couldn't calculate them. So on the train back, we solved the problem, I mean we still had to do some calculations, but we understood it, it was the photodisintegration of the deuteron." We're all laughing; Gerry says that Chadwick shouldn't have bet with these two smart devils, but Hans, still laughing, says that Chadwick probably made the bet so that they would work on it for sure! This is when Hans tells the story of how he was to give a talk about Chadwick's discovery of the neutron in Tübingen in 1933 (which I relayed earlier). The discussion then turns to Gerlach.

"He was a very nice man" Hans maintains, "also Stern."

"Unlike Stark," I interject.

"Stark was a terrible terrible Nazi!" Hans's face grows grim. "Much like Lenard, they were terrible people."

"But Gerlach wasn't nice to Trudi Goldhaber?" Gerry asks.

"No," Hans answers, "they didn't get along well. I don't know what it was, he wasn't nice to her." I inquire about Trudi Goldhaber and find out that she is the wife of Maurice Goldhaber, now at Brookhaven Labs, and was herself at Brookhaven until they "let her retire." She proceeded to accuse the Lab of discrimination which, Gerry admits, was probably true, and was so dismayed at being let go (while her husband stayed on) that she got so sick that she had to stay in the hospital for months, without anybody being able to find anything wrong with her.

"It was all psychological," Gerry shrugs. I find out that Trudi still lives, and that she's doing much better now. Gerry and Betty visit her at times. Trudi Goldhaber is also the mother of Fred Goldhaber of Stony Brook, and the sister-in-law of Gerson Goldhaber, to complete this family of physicists. Trudi Goldhaber is German, Gerry points out.

It is now Gerry's turn to tell a story, and it is spurred by the reminiscence of all these British physicists. It is the story of how he wrote a paper with D. R. Hartree.

"Oh you did?" Hans is surprised.

"Well," Gerry continues, "I had pretty much figured out how to do the Lamb shift in heavy atoms to all orders in perturbation theory, using Green's functions. So all I needed was some wave functions to calculate it, and Hartree had this potential ready, and he was very good on the computer. Well it wasn't really a computer, it was one of these mechanical things where you have to crank, but he was a real wizard with it, so we sat down and he was putting in the potential and doing the calculations for me. Then I wrote the paper for the Royal Academy of Sciences, and in the acknowledgments I wrote 'I thank D. R. Hartree, D.Sc., F.R.S. for doing the calculations for me,' and when Peierls saw the draft he almost gagged, because "you don't thank a fellow of the Royal Society for doing your calculations!"

"But Hartree probably wouldn't have minded, he used to thank his father for doing calculations in his papers," Hans adds.

"Of course not," answers Gerry, "he wouldn't have minded. I think I put Peierls name on the paper too, afterwards."

The discussion shifts toward the drudgery of being on committees, and Hans quips that he has learned from Karman how to deal with these matters: When somebody calls you if you could be on this or that committee, you ask if they want you to work, or whether they only want the name. Then they would stutter somewhat on the phone and then admit that, yes, in fact all they were interested in is to have the name on the committee. Then you agree, and that is that. I ask Hans about (Theodor von) Karman, and he immediately sets out:

"You know that he was here at Caltech, of course." (That is hard to miss, as several buildings here are named after him.)

"In 1940, when Paris fell to the Germans, Teller and I decided that it was time to do something for the war effort. We decided it was best to ask Karman, as he was knowledgeable about these things. So we drove down here ..."

"You drove from Cornell?" Gerry asks.

"No, I was at Stanford at the time," says Hans, and Gerry insists:

"What were you doing there?"

"I was lecturing at a summer school."

"So we drove down here and Karman tells us we should look at shock waves, shocks in a medium, and how it chemically changes as the shock wave passes through. So we looked at it, and it became a very nice paper, but of course we couldn't publish it because it was secret. We found how the medium would become very dense and heat up as the shock passes through it, and, if you have air for example, then first the molecules would get excited and you have some energy absorption from the vibrational modes, but that's not the important effect, the oxygen molecules get dissociated, and then ionized, and NO forms behind the shock. All this eats up a lot of energy so the temperature behind the shock falls rapidly, and we calculated all the observables, the temperature behind the shock and at what distance. It was later published in an obscure place, the "Aberdeen Proving Grounds" in Maryland, because we weren't supposed to put it into a renowned journal." Gerry asks if it is declassified now, and Hans says that it was never really classified, it was just not to be published. Gerry mentions how he was asked to give the hydrodynamics class when he was in Birmingham, when Peierls gave him the notes he himself was using, which were largely Hans Bethe's notes on shock wave formation and propagation, and how he used them over and over again; also when he did the supernova, and how he got together with Hans then and Hans (referring to the supernova) decided to "make the thing explode, come hell or high water." Gerry assures that he still has these notes, in perfect condition, that everything is in there, and starts giving examples, any one of which make Hans nod enthusiastically:

"Adiabatic flow at the beginning of the shock ..."

"Yes, yes!"

"Theory of convection, transfer matrices ..."

"Ah, yes ..."

They lost me quickly in this exchange, as I know absolutely nothing in this field, and therefore can't remember more of the technicalities they were mentioning. Gerry also relays the anecdote where Hans is giving a talk in Seattle about convection currents (these seem to be associated to shocks, one of Hans's preferred subjects) and mentions that a specific equation wasn't directly solvable analytically, and this woman in the audience asks him what he did, and he says that he put it into the form of two coupled differential equations, and solved them numerically. Gerry, grinning now, says he already knew Hans pretty well then and later walked up to him and asked him "How many points did you use" [to discretize the equations] "four or six?" Bethe is laughing out loud "Four of course!," the joke being that Hans never used a computer but did the whole thing on a slide rule! Hans is unbelievable with a slide rule.

We (they, really) talk a little more about the convection currents, and why there is no mixing of material, then we make appointments for breakfast, and the following hike in the mountains. Hans says that it's going to be a slow hike because he can't walk that fast anymore, but Gerry gives me a paper that he received that day and says that I can report on the electric condensate tomorrow during the hike. I take the paper, though I know that Gerry isn't really serious about the report. Also, I didn't have time to look at it since just before dinner, when I had a quick peek, mainly because I had *this* stuff to write up ...

Saturday, January 25th

A long hike in the morning until afternoon, dinner in the evening, and a lot of stories. I wish I could take notes, but of course I can't. I don't think they would talk as freely if they knew I "recorded" everything. For instance, tonight at dinner, Gerry told a story that involved him and C. N. Yang, his direct boss at Stony Brook, in 1967. He said it's not for public knowledge, but here it goes. He (Gerry) was at the time in some way involved with T. D. Lee, either directly or indirectly, I don't remember. It must be said also that although Yang and Lee had the most fruitful time in their career collaborating, obtaining the Nobel prize for discovering CP-violation, they are now in a constant and bitter feud, which, to make it short, mainly was about who would be first author on their joint papers. So Gerry, having something to say about his relation with T. D. Lee, told Yang: "Well, you know how it is to work with him," upon which Yang turned slightly red and asked Gerry into his office. There, he announced quite solemnly and slightly angrily, that "Since I've worked with Fermi, I've never worked with anyone. People worked with me!" Gerry clearly felt reproached by that episode, but adds that he later became one of Yang's best friends, and that he always greatly admired him.

I was at the apartment at 8:30 a.m., for breakfast. We left at 9:00 and drove up Mount Wilson. Slightly below the peak we parked the car and went on a trail that presented a handful of breathtaking vistas of mountains and cities in the mist, which I wish I could have the time to describe, but haven't, as I only have these couple of hours after dinner to write down these notes. I was slightly worried how the hike would go, how Hans would manage it, but as always he surprised me. We left on the trail at 10 in the morning and came back to the car at 3 in the afternoon, and Hans still looked fresh, and quite obviously pleased with his performance. We had set out on the trail on a slightly overcast morning, which made the hiking much easier. Hans walks steadily, with his upper body slightly angled forward and his hands folded behind his back. Wearing sunglasses and a brimmed hat he almost looks disguised, and moves steadily, and certainly not hesitatingly. As I still have sore muscles from the workout the day before, I'm glad he is setting the pace. So we walk along the trail, over difficult terrain, up and down, large sections through treacherous snow that covers the shaded sides of the mountains. Wearing only sneakers I almost slipped a couple of times, while walking about a meter behind Hans, hoping to catch him should he fall. Of course, he never did or came close to it (unlike I), not to mention that I probably couldn't have held his weight. We do not talk much on the way down to "inspiration point" where we would eat, rest and then return. When we did talk we chatted a bit about his family, of his daughter who is married to a Japanese man, has two kids and lives north of Kyoto in the mountains, and his son Henry and grandson Paul, named after Paul Ewald, his father-in-law.

Hans also talks about how his children learned German after the war, vacationing near Munich in a house or mansion that once belonged to Rose's grandmother, and that was returned to them after the war. Gerry and I also discuss some physics, but I'm not really interested, so Gerry quits after a while. On the way back we discuss school, after Gerry complained that his son Titus never does school work, and complains about me having given him a bad example by telling him that I hadn't either.[e] I defend myself by saying that I told him also that you have to work hard at least once in your life, and that I did so in the first years at Stony Brook. Hans thinks back and notes that he worked hard from age 13 to 16 learning calculus, thus just after the first war, and that that gave him enough of an advantage for the rest of his school years. I continue my thread telling about how I took my qualifying exams after two years (trying to impress Hans all at the same time, of course) and when Gerry asks whether I passed the exams as the best of the year, I told him that I did pretty well, but wasn't the best, and as defense mentioned that I didn't have the time to study more than a month, being in the middle of finishing a paper. Asking me about what paper that was, we come to talk about my charmonium paper[f] again, and since I smell a way of getting some stories out of Hans again, I tell him again how I only had to somewhat change the differential equations, which I got out of his book. So he turns to me and asks:

"So you think I had already done everything, then?" Naturally I say yes, which in fact is pretty close to the truth, at least concerning that paper, and he smiles and says: "That is nice to know." I then go on to ask him about Cornelius Lanczos, who wrote the original papers about the strong-field Stark effect, only to find out that he never read these papers (although the results appear in his book). Immediately then I turn to Schrödinger's original papers about the low-field Stark effect (which I had read at the time) and of course, those he had also read. We take turns marveling at some details of these really very nice papers, and go on to Schrödinger's personal life, as is inevitable. Schrödinger was, as supposedly everyone knows, a compulsive womanizer. Hans maintains that he in fact kept several mistresses at the same time, while all the time not only his colleagues, but also his wife, knew all about it. Hans describes how despite everything she always would stick to him, but would finally take a lover herself, which we found, after all, was just fair. Because of his unstable life Schrödinger had difficulties finding a permanent position after he quit the University of Wien, and ended up in

---

[e]Titus is now finishing his Ph.D. in biology at Caltech.
[f]*Phys. Lett. B* **217**, 5 (1989).

Ireland, from where he would take frequent trips to the States. Strangely enough, Hans remarks, Schrödinger never achieved anything of similar importance after his series of papers on quantum mechanics. He didn't do anything worthwhile before, and after that he was so famous he thought it better to write more philosophical papers. The Schrödingers, as it turns out, were close friends of Hans's parents-in-law, the Ewalds.

We talk briefly about Kramers, whom Hans genuinely admires, touch on Nazi war atrocities against the Dutch (Kramers being Dutch) and somehow end up talking about the first neutron time-of-flight detector made by Bob Bacher, who was later in the Atomic Energy Commission. Hans mentions that before that Bacher was the leader of the experimental division in Los Alamos, and that for me is the signal to ask about his Los Alamos recollections. Hans, of course, was the leader of the theoretical division. We talk briefly about the general structure of the Los Alamos project, which consisted of the divisions just mentioned and the engineering division. Each division had a number of groups, with group leaders coordinating the efforts. Somehow we end up talking about the calculational effort, and I ask Hans what Feynman's contribution was. Hans maintains that his contribution was enormous. He was named the group leader for computation after a while by Hans, and singlehandedly led the team to success. In the beginning, they were doing their computations on hand-held mechanical machines, a tedious and time-consuming task. Then someone suggested that they should get "one of those IBMs." Of course, money, or any resource for that matter, was no object, and within a couple of days a great many crates arrived, with minimal unpacking instructions. This proved to be the challenge that Feynman, and his assistant Eldred Nelson, needed. Of course they knew nothing about computers to start with. Nonetheless, they had the machines running after slightly more than a week. Ten days after delivery, an IBM specialist was sent to set up the machine and just couldn't believe what he saw. The technician conceded that this had never been done before. After that, the task began to program the machines and obtain results. Millions of cards were punched and filed, and results were pouring in one by one. All the while, Feynman was supervising the effort and coordinating the calculations. (His account of that is in his book.[g]) I remark that it's strange that a man of Feynman's abilities would be overseeing the numerical activities, and Hans answers that this was a critical piece in the bomb effort, and that Feynman had a great talent for things mechanical. Besides, he says, Feynman could do absolutely anything. He further tells the story that earlier, the group

---

[g] R. P. Feynman, *Surely You're Joking, Mr. Feynman!*, W.W. Norton, New York, 1985.

leader in computation was a young physicist by the name of Frankel.[h] He was very much in love with these machines, and would play with them all day, which Hans said he could understand. Nevertheless, one day he decided that "this will not do," and asked Feynman to take over, giving Frankel to Oppenheimer. At the same time, he ordered another young physicist to assist Feynman, by the name of Metropolis,[i] who, as Hans puts it, "took to the computer as a duck to water." With Feynman and Metropolis on the program, the computational problems were solved. I cannot help but feel cheated that Feynman died so early. If he hadn't, he might possibly be walking right alongside of us now, and I could listen to his Brooklyn accent.

While talking about the group leaders, the name Klaus Fuchs falls. I inquire about him. "He was one of our most valuable group leaders" Hans comments. I want to know more and ask whether the damage done by Fuchs was important.

"He gave them everything, on a platter," he answers, very slightly animated. "He knew all the details, knew them better than I. He knew how to make the plutonium, and he knew about the implosion. We calculated that the Russians could make a bomb in an absolute minimum of five years, starting in 1945. They tested their first bomb in 1949, after four years! But Fuchs had started giving them everything we had since 1942! He had an agent in Santa Fe that he met regularly. All the while he was a very important, and very effective group leader." I ask what became of him, and learn that he was the deputy director of the (now defunct) Nuclear Research Center Rossendorf near Dresden, in East Germany. "He never regretted what he did," Hans says, shaking his head, "and that, I resent." I ask why Fermi was not in Los Alamos during the time, and Hans answers that he was, during the end, but that he was needed more in Chicago (Fermi lectured at the University of Chicago) where they were building and testing the reactor that was supposed to breed the plutonium. The pile was being built by the Dupont company, and Fermi was supervising their efforts. Hans tells of one event where the reactor was finally working satisfactorily, and Dupont decided to shut it down for a while. When they tried to activate it again, it wouldn't. The reaction just died every time they tried. So they called "Mr. Farmer," which was Fermi's secret identity. (Of course we asked Hans what his secret identity was, but he said he had none, that only when he was travelling for the project he went by the name of "John Doe.")

---

[h]Stanley P. Frankel, then postdoctoral fellow of Robert Oppenheimer.
[i]Nicolas C. Metropolis (1915–1999), co-inventor of the Monte Carlo method.

Fermi started to look at the problem. After a couple of days of intense testing and thinking, he came up with the answer: Among the fission products generated in the operation, it turned out that a Xenon isotope had formed, "with an immense neutron capture cross section," which would stop the chain reaction. Once they figured this out, they could restart the reactor. Hans maintains that without Fermi's intervention it would have taken them weeks to figure out what was wrong. Gerry asks whether Niels Bohr had any part in the effort. Hans smiles:

"Yes he did, an important part. It was in 1944, we were looking for an initiator that would start the implosion, to get all the uranium, or plutonium, in this case it was plutonium, close together. We needed something that would work fast, and reliably. The problem was to get enough neutrons quickly, but there should be absolutely none before the initiation. We put two groups to work on the problem, supervised by three people, Critchfield, Fermi, and myself. Fermi was heading one group, Critchfield and I the other. Fermi came up with a design, and so did Critchfield and I, and we went into testing. It turned out that the Critchfield design worked really well, whereas Fermi's was not a hundred percent. Nevertheless it came to a vote, and Critchfield and I voted for our design, while Fermi voted for his. However, it wasn't very wise to contradict Fermi. So we asked Niels Bohr to look at the test data. After two weeks he came back and recommended our design, which settled the matter. So this was Bohr's contribution: you circumvent higher authority by going to an even higher authority!"

At dinner we had several discussions, of which I will pick out only one. Because it seemed appropriate at the moment, I asked Hans about Majorana, the famous Italian physicist who disappeared while still very young. He is presumed to have committed suicide. I ask him whether he met him.

"Indeed I did," he answers, "as did everybody. He was a very shy man. I couldn't communicate with him much, as he spoke neither German nor English, so we could only communicate in my poor Italian."

"It was in Rome in 1931–1932, all of the Italian physicists were there; Majorana, Segrè, who was Majorana's mentor, Amaldi, Fermi and Rasetti. Of them, Majorana was clearly the brightest." He briefly relates the story of how Heisenberg had constructed a theory of the neutron. Majorana read it, realized that it was wrong, and proceeded to construct the correct theory.

"Then," Hans continues, "there was the problem of getting everybody a position. The trouble was that there was one postdoc, who clearly was the worst of the bunch, who was the son of a very important politician. To win a position, however, everybody had to enter the 'concorso,' and it was clear that Majorana would win. There were three openings that year, and in the end Majorana got the position without taking the concorso, Amaldi came in second, and somehow the politician's son got the third position." There is a brief silence, and Hans's face turns grave.

"I think Majorana committed suicide. He was a very silent person, very insecure, and he only confided to Segrè. When he got the position, he had to teach, and he was very uncomfortable with that. He didn't like to speak in front of an audience, he was so insecure."

During the discussion I found out that Hans and Fermi were speaking German to each other, just as Hans and Segrè. I also found out that Mrs. Peierls had a very loud voice. Many more things were discussed during the hike and at dinner, which I am now too tired to remember or to write down.

Sunday, January 26th

Dinner at Hans's and Gerry's tonight, with two guests, Judith and David Goodstein. She is archivist and faculty associate at the department of history here at Caltech, and just published a book about Caltech's history.[j] He is professor of physics and vice provost, also here. They arrive at 7 p.m., just after the Super Bowl game has ended (the Redskins won against the Bills, 37–24). We have hors d'oeuvres in the living room (which also is the dining room, but we sit around the couch table). The conversation is good-natured, Hans is the center of attention, naturally. The Goodsteins had both known Sir Rudolf and Lady Peierls, and stories about her abound. Gerry imitates her accent: "Zis how you clean pans, Bethe?" and Hans laughs loudly and cheerfully. Hans has a very "German" laugh, it is really a loud: "Ha, Ha, Ha, Ha." He proceeds to tell the story of their important dinner, at the house of Peierls in 1934 in Manchester.

Peierls and Hans were very young (Hans was 28, Peierls a year younger) and they had an important Italian physicist as a guest. Mrs. Peierls had cooked a duck, and apparently it turned out pretty tough, so Mrs. Peierls called Hans into the kitchen to help with the carving. The job turned out to require considerable strength, and all the while that Rudolf Peierls was entertaining the guest, Hans and Mrs. Peierls were finding themselves pulling at a duck's leg each, leaning against it to the point of rupture, as Hans and Mrs. Peierls were heard flying through the kitchen. Hans has a marvellous way of relaying this story, with a keen sense of comedy, and we're all laughing

---
[j] J. R. Goodstein, *Millikan's School*, W.W. Norton, New York, 1991.

with tears in our eyes. Next in line is Hans's story about how his son got hired at Chase Manhattan Bank. In fact, as he recounts, what Henry was really good at, is Bridge. Gerry imitates Mrs. Peierls again: "Boy need only be good at one thing!" To our amusement Gerry maintains that Lady Peierls decided what would be the future of both the boys he then had (she was not interested in the fate of girls), and of many other professor too.

Hans continues his story. In the New York Bridge Club his son made the acquaintance of a Chase Manhattan executive, who took a liking to him, and one day mentioned that his bank is looking for a young, bright person, and motivated Henry to present himself. So he did, and was interviewed by some people from the personnel department.

"In what branch did you obtain your Ph.D.?" he is asked, and to their astonishment answers that he does not have a Ph.D.

"Dear Sir," he is lectured, "we are turning away Economics Ph.D.s by the dozen, how do you expect to get hired for this position? What degree do you have anyway?"

"None," Henry Bethe answers, and is asked to leave. He relays this to his friend at the Bridge Club, who arranges another interview, this time by the Bankers themselves. They ask different questions. They ask what he would do in specific situations, like what he would do if their Japanese branch would be losing money, and such. After an hour of interview, he is hired. This leads to an internal investigation of hiring practices, which leads to nothing. When they review his performance after one year, they relay his excellent performance to a higher executive of the bank.[k] He looks at the report and asks: "What degree does this Henry Bethe have?" The reporting person shuffles his feet nervously:

"None, Sir."

"Well, then get him one!"

So they go to Henry Bethe and present him with a choice: "Do you want us to buy you a degree or do you actually want to work for one?" Choosing the latter option he obtained an M.B.A. at Columbia University, while working at Chase.

We go to sit at the table. The discussion is mainly politics, specifically the Democratic bid for the presidency, which we all agree can only be successful in 1996, possibly with Cuomo as candidate. Hans, while liking Cuomo seems to favour Lloyd Bentsen of Texas, as he believes that the Democrats "have to carry the South." We also discuss the disintegration of the Soviet Union.

---

[k] I learned later that this "higher executive" was none other than Nelson Rockefeller.

David Goodstein voices his fears that the world is now a more dangerous place, but Hans disagrees.

A snippet of interesting information surfaces. Hans was, as is well known, one of the chief advocates and negotiators of a treaty banning nuclear explosion testing (during the Khrushchev era). This treaty was near completion when all negotiations over it were unilaterally cut off by the Soviet Union. The incident that had apparently provoked it was an unauthorized spy mission by a U2 spy plane over the Soviet Union, unauthorized by the latter, of course. Recently, however, Hans talked to one of the negotiators, to whom he still has contact. He didn't reveal his name, but hinted at the fact that he is, or will be, the ambassador to Sweden. I suspect it is (Georgi) Arbatov, whom Hans had mentioned some days earlier, and to whom he had given the letter to Yeltsin.[1] In the course of the discussion with this official, Hans remarks that he was truly sorry about the U2 incident, and that that had prevented the signing of the treaty. "But no," the official exclaimed, "this was not the reason at all!" He then went on to describe a sudden shift in the power structures of the Soviet Union, completely undetected by Western observers, that forced Khrushchev to take their hard line. The U2 incident was a welcome excuse to not sign the treaty, and subsequent echoes of the hardliners' influence where the Cuban missile crisis (which, as is now revealed, was much more serious than believed at the time, since live nuclear missiles were actually stacked in Cuba), the building of the Berlin wall, and the ousting of Khrushchev himself.

Judith Goodstein has a new book project, and as it is a collection of Feynman's writings, a "Feynman Reader" of sorts, she is asking for our opinion about several choices she made. We generally agree that her choices are interesting. I propose to her to use excerpts of his 1968 *Acta Physica Polonica* article on his attempts at quantizing gravity and Yang-Mills theories, and his discovery of ghosts, as it is really a transcript of a recording that was made while he was giving a talk at a Polish summer school. I was pretty sure that she didn't know the article, and indeed nobody does, not even Hans. But it is the most "Feynman" article I ever read; it is amusing and sobering and just about awe-inspiring. At the table, I paraphrase some of it and everybody loves it. Ending my little discourse, I feel a little saddened again. The table we are sitting at seats six. To my right is Hans, Gerry is opposite of me. David Goodstein sits opposite to Hans, whereas his wife Judith sort of heads the table, between her husband and Hans. The

---

[1] Instead it turned out to be Oleg Grinevsky, former security adviser to Khrushchev, and ambassador to Sweden from 1991–1997.

place on the opposing end of the table facing Mrs. Goodstein is empty, and this is precisely the place where Feynman would sit now, had he lived, as he was a good friend of all of those present, excluding me. I mention this and David Goodstein agrees, yes, he would probably sit there.

Mrs. Goodstein seizes the moment to ask Hans whether he could date a letter that he once sent to Feynman, and that she found in the archives. She describes the contents, namely Hans urging Feynman to write up what he had, to "publish now, before someone else does. It is time to write things down!" Hans smiles:

"This was good advice wasn't it?"

"So you do remember the letter?" Mrs. Goodstein asks. Hans has this expression on his face that I have come to associate with moments where he is extremely pleased with himself.

"Indeed I do," he answers, his eyes beaming out of slits, with a smile which I can only describe with the German word "verschmitzt."[m]

"If I give you a copy of the letter could you date it then?"

"I can now, if you want!" he triumphs.

"I was on sabbatical leave from Cornell at Columbia University, it was 1948, the spring of 1948!"

Hans's memory is, as must trickle through these pages, absolutely phenomenal. Subsequently he reveals that he keeps a diary up to this day, ever since his Los Alamos days, after he found out from Robert Bacher (the head of the experimental division at Los Alamos) that he was keeping one, a very accurate one. Bacher's Los Alamos diary turned out to be invaluable, as the secretaries there strangely enough did not record the daily goings-on. It is, however, still classified. Someone asserts that most of what is in there would probably be known by now. "Yes," Hans answers, "most of it, except the dates!"

The discussion turns to J. R. Oppenheimer. This is a dicey topic, as Gerry and Hans cannot agree, and decided some time ago that they would not discuss Oppenheimer anymore. Nevertheless, everybody wants to know their impressions. Gerry starts by saying that Oppenheimer is Hans's hero. I look over to Hans, and his face literally says, "Yes, he's my hero." Hans says that nobody but Oppenheimer could have led the Manhattan project, and that he did it absolutely brilliantly. Gerry, being a postwar physicist not involved in the project, only knows Oppenheimer from his time in Princeton, when Oppenheimer was the director of the Institute for Advanced Study, and he a professor at the University. There, Oppenheimer must have been vicious,

---

[m]Roughly: "mischievous, whimsical."

especially to younger people. Of an allotted 90 minutes in a talk, most speakers would not be able to hold 10, as Oppenheimer would inevitably interrupt. Hans interjects that he managed 30 minutes, which everybody in Princeton who knew Oppenheimer characterized as magical. Gerry however goes further (he had a self-admitted 7-minute session once). He maintains that Oppenheimer would first take out the speaker to lunch, ask him about his topic in detail, only to use it against the speaker later. After having made such a point, he would turn to the audience and sneer. This particular habit made Oppenheimer impossible in Gerry's eyes. Hans tries to defend Oppenheimer by citing the fact that he wasn't treated very well by the authorities. Gerry counters with his experiences with the "House Committee on Unamerican Activities" (the McCarthy investigations). While Gerry was in Birmingham, McCarthy found out about his activities in the Communist party while a graduate student at Yale. They asked him to denounce his comrades (some of whom obviously had given away Gerry), which he refused. Instead, he gave the letter from HUAC to Professor Peierls, who placed it in a safe, and went on with his business. Gerry was denied a passport for the next seven years, which forced him to stay in Birmingham. (This turned out to be lucky in retrospect, because Birmingham was such a good place under Peierls.) Mentioning this, Gerry said that *that* didn't make him treat younger students badly, looking all the while at me. Hans, however, while acknowledging Oppenheimer's misbehaving, still maintains that the project was unthinkable without him. Yes, maybe Fermi could have led it, he concedes, but Fermi was out of the question because of his being Italian. Mrs. Goodstein asks Hans if he thinks that Los Alamos was Oppenheimer's finest hour. Hans is quite solemn:

"Yes, this was his finest hour."

She further asks Hans about a movie about Los Alamos, "Fat Man and Little Boy," starring Paul Newman as General Leslie Groves.

"That was a terrible movie," Hans answers, "it was wrong everywhere."

"They started with the premise that Groves forced us to do the project. This wasn't true at all. We had thought about it, and we wanted to do it, we had pretty much figured out what to do before Groves even came to the project."

Hans proceeds to tell the story of how the movie producers asked him to use his character in the movie. They sent him the lines attributed to him, and as there weren't many, he agreed. But when they sent him the contract, which forbade him ever to make any use of any material related to the events in the movie, or to make any other deals with other movie companies, or to write about it, he flatly refused to sign. They sent the contract back, having added a lot more lines to his character. He wrote back that he didn't feel portrayed accurately, that he wouldn't stand for the lines written, and that they should attribute them to some unknown character. That was the end of the Hans Bethe character in that movie. However, he says he immensely liked the series that was done for PBS. He liked the way they portrayed him, and especially he said that the actor playing Oppenheimer seemed more like Oppenheimer than Oppenheimer himself!

The last discussion of the evening centered around the former Soviet Union again, with everybody relaying some tidbit illustrating glasnost, and Gerry blaming the US government for not realizing what's at stake, and sending the wrong signals by shipping to Russia only surpluses that weren't needed, as well as juice that turned sour and bread that was stale.

I have to admit that during the dinner I very often felt the urge to interrupt, or at least make my opinion on this or that be known. The hardest thing however was *not* to relate some stories that I knew, and that I figured would fit perfectly, in ordinary circumstances. However, I also realized that Hans was center-stage here, not I, so I tried to hold back. I "broke down" only at the end, telling three stories in a row, on the recent revelations about the Soviet moon program, the delicate matter of how a DEC computer's central mother board would exactly fit into a Soviet-made computer (while strict export limitations prohibited US computer parts to enter the Soviet Union), and of the friendly American visit to the Soviet Mission Control room in preparation for the Apollo-Soyuz mission, which ended in an éclat because the Americans simply did not believe that the equipment they were shown could guide rockets and satellites, and assumed that the Soviets kept secrets. While all these stories were very well received, especially Hans exclaimed many times: "I didn't know that!", "This is fantastic!", I feel a bit bad about it now. Gerry walked me home, and in a joking manner alluded to my "staging myself," though acknowledging that I did give Hans the opportunity to talk (Gerry likes to exaggerate). I assured him that I was well aware of who was center stage, and that I tried very hard. I did, really.

Second Week

Monday, January 27th

Nothing to report today. We had lunch at the Athenaeum at noon, the usual troika. From the conversation, nothing seems to stand out. I told some gossip from Stony Brook, and we compared sizes of quantum mechanics classes. My first year at Bonn University had about a hundred students to start with, only to be narrowed down to about sixty towards the end. I told Hans that in some universities in Germany there could be over 500 students, which he found incredible. Gerry asked how big his class was. Hans thinks for a second, and then says, "Twelve." Of course, *his* class was taught by Arnold Sommerfeld. Gerry asks if he was the best in his class. Hans answers that this is a hard question to answer, meaning, I realized a bit later, that although it was he, he didn't want to say it. He only mentions (this was in Munich by the way) that as long as Rudolf Peierls was in the same class, this question could probably not be decided. But, he said, Peierls deserted them a year later. He continued his quantum mechanics at the University of Leipzig. Werner Heisenberg was lecturing there.

No dinner tonight, as Gerry and Hans were invited to dinner by Stanley Sheinbaum, who from what I could gather is a wheeler and dealer in the Democratic Party. So I took off to work out. Tomorrow Gerry wants me to give a talk to Hans about the paper we're currently writing. I just wrung some plots out of the Stony Brook computer for that purpose.

Tuesday, January 28th

I gave part of the talk before the lunch seminar, part of it after. I sort of tried to make it self-contained, without too many details, but I underestimated Hans. He just simply understood what I was doing, asked the usual pertinent questions ("well, yes," I would answer, "actually I hadn't thought about this, but it's a good question ...") and was very nice to me about everything.

At six I went over to Hans and Gerry because they were watching the "State of the Union" address by Bush. There was another guest watching, who would have dinner with us. It turned out to be none other than Bob Bacher, whom I've mentioned earlier a couple of times! He was the leader of the experimental division in Los Alamos (and formerly provost of Caltech). So I had dinner with the leaders of both the theoretical and the experimental division, at the same time! I had to remark to Hans later that this could only be beat by also having Oppenheimer, but of course he is dead. Bacher is a very nice man, he looks a bit younger than Hans but is actually a year older. He lives in a retirement home with his wife, who is ill. Of course, during dinner we discussed mostly politics, which is why this report won't be very long, and while Gerry was away on the phone and Hans and Bacher talking to each other, I couldn't help thinking what a waste it is that these two brains will, some time in the future, just simply cease to exist, and all of their knowledge and experience that it took 86 years (up until now!) to accumulate, could then not be put to use anymore. The political discussions were mostly on Bush's announced measures for economic recovery, and military cut-backs.

Wednesday, January 29th

High noon: lunch at the Athenaeum. A hot, slightly hazy day. While eating, we talk about books, plays, and literature. We talk about several books that Betty Brown once gave me to read, and which Gerry has all read, it turns out. One that Gerry liked very much is Jorge Amado's "Dona Flor and Her Two Husbands." We tell Hans the story, who finds it rather amusing, and then starts to recite two passages from (Kurt Weill's) "Die Dreigroschenoper" in German, dealing with very much the same amorous subject matter as the book we told him about. Hans has no accent in German whatsoever. He very rarely speaks German, only when quoting, or to use a word that is best illustrated in German. Never for conversation. Hans wants to know what kinds of books I had to read for my German class in school, and how the exams where held. We very briefly discuss Döblin's "Berlin, Alexanderplatz," Fontane's "Effi Briest" (Hans didn't read that one, only heard of it) and Flaubert's "Madame Bovary." We also bring up Dürrenmatt (Hans mentioned him), and Gerry mentions "Die Physiker," which Hans immediately characterizes as "a bad play." Gerry claims that the character 'Möbius' in the play is in reality Konrad Bleuler, while I of course claim that this would be the character "Beutler," and Hans agrees. Gerry mentions that Bleuler and Dürrenmatt were neighbors for quite some time.

We come to speak about the *Physics Reports* volume that was edited on the occasion of Hans's 80th birthday celebration,[n] and held at Cornell University five years ago. As an epigraph, it has the following sentence:

> "Das sind richtige Männer, die wacker die rauhen Berge besteigen.[o]"

As there is no attribution, I had always wondered where this phrase came from. The origin is actually quite amusing. In fact, this was said by a little Swiss boy walking with his mother, after seeing Hans and friends enter a mountain pass in the Swiss alps, all dressed for a long walk. This is

---

[n] *Phys. Reports* **163**, 3 (1988).
[o] "These are real men, who fearlessly climb these harsh mountains."

very hard to imagine, but a true story nevertheless. The boy probably has no idea that, and in which way, he was immortalized.

Dinner at 6:30. Surprise of the day (besides the fact that CERN couldn't offer me a position) is that Hans got a phone call announcing that he won the "Einstein Peace Prize," an award worth $25,000: not bad. He seems to be quite happy about this, as he didn't expect it, and says that it is nice that his other work also gets recognized. More discussion about Bush's address yesterday follows, and this time we have it about education, about how bad it is in the States, and compare it to other systems. I compare with the German and French school systems, Gerry knows a lot about the English one, as he spent ten years in England. We all agree on the main points: more money for more schools and more teachers with a higher salary. We also agree that it'll take many, many years. Gerry mentions as exceptions a Science School in the Bronx, which he claims has fostered more Nobel prize winners than all of Germany. While this claim is of course preposterous, I ask him who came from there. He can only name two, in fact, but they are quite illustrious: Steven Weinberg and Sheldon Glashow. Moreover, they were classmates, in the same grade!

We continue a discussion we had at lunch, about school. Hans is interested in how history is taught these days, this subject came up some days ago already. Especially he says that he wasn't taught about any other country than Germany, except for those that Germany had fought a war with. He remembers that they weren't taught anything about the United States, they knew that it was discovered once, and what it was called. I reply that we weren't taught about the history of the United States either up until the last two years, during which we covered mainly 20th-century history. He then asks me how religion was taught. I told him that it was compulsory up until the year where I was in fact expelled from religion class, a story that he wants to hear. So I tell him about my belligerent last year in religion class, where I thought that it was my duty to tell everybody what fools they were, believing all this nonsense (I was fourteen). Hans is clearly pleased, and laughs out loud when I tell the story where one "faithful believer" tried to contradict me all the time in class, then reverted to asking the question that was supposed to deliver the final blow: "But how was this universe created then?," to which I, slightly heated from the discussion, yet still calmly, answered that I would gladly proceed to tell her just that, which I attempted, only to be thrown out of class a little while later because I was "disrupting" it. As fantastic as it sounds, it's actually a true story. The teacher by the way didn't throw me out out of malice,

I got along quite well with him, it was just because I was "not what was needed."

Hans likes all that, he tells us that he felt pretty much the same at pretty much the same age, but that he very rarely aired these convictions, but rather kept them to himself. Hans asks Gerry if he ever was as combative, and while I understood this question to relate to matters of religion, Gerry answers generally, by saying that he mellowed out at age 40–50. He goes on to tell about the fights he had with Aage Bohr and Ben Mottelson at Nordita, the Danish Research Institute founded by Niels Bohr, in Copenhagen. Gerry spent many years there upon invitation by Niels Bohr. Gerry feels that the nuclear physics division at Nordita is in serious decline, mostly due to the fact that, after Ben Mottelson and Aage Bohr (the son of Niels) had received the Nobel prize, everybody there just worked on extending their theory.

Hans is wondering what Edward Teller might think about all the nuclear disarmament going on in the world. Hans, a strong proponent for nuclear disarmament (his engagement in these matters having won him the Einstein Peace Prize) is "pleased as pie," as Gerry says, every morning reading the headlines announcing further and further cuts. Hans also maintains that Teller is a real nice guy, if you don't talk politics with him. As Hans is going to Stanford next week where he is going to meet his good friend Sidney Drell, another disarmament proponent, he will try to find out.

Leaving for the night I wish Hans a good night. He's sitting at the table reading *The New York Times*, looks up, and says: "See you later, alligator!"

Thursday, January 30th

On the way to lunch Hans makes a few comments about the subject of my little talk on Tuesday, and I'm sort of pleased that he takes some interest in it. The day before, on the way back from lunch, I asked him explicitly what he thought about the idea of using the ρ meson mass instead of the quark condensate as the order parameter of the QCD phase transition. As I cooked that idea up on the plane coming here I didn't feel too sure about it. But Hans said that that was quite all right, and that I had basically proven that it didn't really matter, hadn't I? That was indeed the point, and thus I'll publish it.[p]

Very quickly at lunch we're back in the postwar era, with Gerry retelling how he went to Yale on the G.I. bill after serving in the Navy. He compares his $4 (weekly) room, and $1 lunches to our times, remarking that "not

---

[p] See *Phys. Rev.* **D46**, 478 (1992).

everybody has as expensive tastes as you," into my direction. I'm of course used to this kind of snide remarks; Gerry just loves to make them. Hans throws in an experience he had with Rose on Mt. Rainier in Seattle, "in 1937" as he recalls without a moment's hesitation.

"We were stopping at the 'Paradise Mountain Lodge,' and for $1.50 you could eat whatever, and how much you wanted. And it was very good food. We had fish, and then some other dish, and then two desserts!" Gerry goes on about his first year at Yale, and how his mother had given him $100 to buy himself a suit when he went to Yale Graduate School. He instead spent $20 to buy a suit at the second-hand shop, which was run by a person who bought the clothes from the students who had wealthy parents, a common picture, at least at Yale. Hans asks if the suit fit. "Roughly," Gerry answers. I ask what he did with the rest of the money. "Probably lent it to my leftist friends," is the reply. This leads into a discussion of McCarthyism, and I ask both Gerry and Hans who or what had finally silenced McCarthy.

Gerry describes the methods and motives of McCarthy and his two assistants, Cohn and Schein, the latter being, as Gerry points out, "a homosexual." I wasn't about to correct his substantivation of an adjective. Then Schein got drafted into the Army, and suddenly McCarthy focused his wrath on "uncovering" communists in the Army. Everyday at 5:15 in the afternoon, 15 minutes before the newspapers had to finish the stories for the evening editions, he would stand on the steps of his senatorial mansion (or wherever he made his declarations) waving a pile of papers and saying: "I discovered 27 communists in the Defense Department." He even accused General (and former secretary of State) George C. Marshall, the highest ranking military. After this episode, and the counterattack of the military, McCarthy had sort of lost his credibility.

Gerry recounts how he left the country in the beginning of 1950, more because he wanted to escape his advisor Gregory Breit, than because of the uncertain political times (McCarthy had just started his witch hunt). He wrote to universities in England, and from Blackett got the response that they had no jobs, from Max Born that he (Born) was too old to take on assistants, and only Peierls answered "Come on over!" From another physicist at Yale Gerry was told that in Birmingham he should also look for a certain "Klaus Fuchs" who was a good friend of Peierls. At the dinner reception at the Peierls' house Gerry promptly asked Peierls where he could find a physicist by the name of "Klaus," the last name having escaped him. According to Gerry's testimony Peierls "turned green," then answered that he just visited him in prison, as he was arrested days ago. Gerry, meeting

Peierls for the first time, thought it better to change subject, as the last one somehow had lost its appropriateness for polite dinner conversation. As it turned out, Fuchs was indeed arrested a few days earlier (Gerry arrived February 4th, 1950) for espionage. Hans observes that this was especially hard for Peierls, as he was probably the only close friend of Fuchs (in response to my inquiry, Hans confirms that Fuchs was a loner), and somehow ended up watching out over him. Peierls admitted that he felt personally betrayed by Fuchs. He spent 12 years in prison, teaching physics there, and relativity to his cell mate. A prison guard, or was it the director, Hans isn't sure anymore, once asked Fuchs how this was possible. He answered, "But you are much smarter than my cell mate, I could teach *you* in no time!" After the twelve years, he went to East Germany, and became deputy director of the Nuclear Research Institute in Rossendorf, the biggest one in East Germany. Gerry visited the Institute some years ago, and had lunch with Fuchs. I do not know the year in which Fuchs died,[q] but Gerry remarks that he always had low blood pressure.

Further information about Mrs. Peierls surfaces. She is Russian, "Eugenia" by first name, "Genia" with a soft g, to friends (that solves the mystery about the terrible accent that Gerry always uses when imitating Mrs. Peierls). Gerry describes the extremely nosy and prying nature of Genia Peierls, how she was cornering everybody until she had extracted the most private and intimate secrets, then proceeded to offer her advice. Her advice, Gerry concedes, was almost always sound, and opened up avenues the advisees often hadn't even thought about. He offers the example of his son, who was "ordered" by Mrs. Peierls to go into the hotel business after she had analyzed his personality. He is now running several large hotels, one of them, I was told, in Santenay, France. Gerry asks Hans whether the inquisition staged by Mrs. Peierls had annoyed him.

"No," he answers, "I rather enjoyed that. It was her loud voice that sometimes annoyed me." Gerry asks about how Rose felt, and we learn that she couldn't stand her. After all Genia Peierls had the habit of telling the professors' wives that they should stay at home and care for their great husbands. (This was also observed by Judith Goodstein some days ago, who also was offered this unsolicited "advice.") Hans and Gerry agree however that Genia Peierls was the most exceptional woman they ever met. Peierls had met her on a conference in Odessa, and married her very soon after.

Gerry recounts how Genia decided to marry Rudolf. The latter could sleep in any situation. Once they travelled to Kiev by train, and Rudolf

---

[q]January 28, 1988.

Pcierls, being fairly short, crawls into one of the overhead luggage nets, and falls asleep. The hyperactive Genia sees this and decides on the spot: "This very stable man. This is man for me!" They didn't spend a lot of time in Russia. But apparently Peierls was very good at languages: the Peierls conversed in Russian with each other.

Someone passes our table and greets Hans enthusiastically. Gerry immediately proceeds to tell that Hans got the Einstein Peace Prize, all the while Hans is trying unsuccessfully to prevent that by waving his hands.

"They didn't tell me on the phone to keep it secret," Hans pleads, "but maybe we should wait until they had a chance to make it public." Gerry reluctantly agrees. I ask Hans whether he knew that he would be up for the prize and he denies it, says instead that it came as a complete surprise.

We come to talk about the Nobel prize, and I recount how Einstein was nominated five times, and got turned down four times, mainly due to the influence of H. A. Lorentz, who claimed that Einstein shouldn't get all the fame for special relativity, and later arguing that general relativity was too speculative. They finally gave it to him for the photoelectric effect of 1905 (I heard this version of events from Max Dresden[r]). Hans is surprised. Then Gerry starts to tell the story about how Hans heard about his Nobel prize, in 1967. Gerry always does that, starting stories for Hans. Hans interrupts pretty quickly, saying that he tells it all wrong.

"That night," he begins, "Rose slept in another room. At six in the morning the telephone rings, and they tell me that I won the prize."

"Even you were too excited then to go back to sleep, weren't you?" Gerry jokes.

"No, no, but the telephone didn't stop ringing! The radio, television, newspapers called. This went on until eight. Then finally Rose woke up and asked what this was all about, why the telephone kept ringing. At that moment it rang again and she picked it up. It was the (Cornell University) physics department, and they suggested that they should give a press conference."

"You might as well," Rose answered, "the day is shot anyway!" This, as Hans remarks, was one of her famous one-liners.

Wolfgang Pauli's remark about Peierls (told by Hans): "Peierls thinks so fast, when I finally start to understand what he is telling me, he is already explaining to me why it is wrong."

Dinner. We have roast beef, as this is one of Hans's favorite. We talk a lot about history, especially the French Revolution, the Napoleonic wars, and

---

[r]Professor of physics at Stony Brook.

the war of 1812, between the United States and England. The subject came up when we tried to date "Pride and Prejudice" by Jane Austen, which Gerry is currently reading. Hans supposed "beginning of 19th century, Napoleonic wars." It turned out to be published in 1813. Can't beat the guy. He remarks how the Austen family was always poor, despite being quite influential, and two of Jane's brothers being admirals. Hans loves to talk about history, so during the conversation I try to wring out of my brain what I can possibly find.

During dessert, I ask Hans whether he knew Einstein. He says he met him, but didn't work with him. He also says that he appeared to be a very sweet old man. They met in Princeton, on account of some peace initiative. I ask him whether he ever wanted to work with him, and he answers that he was no expert in Einstein's field, and also didn't want to. At the time, Einstein was involved with his Unified Field Theory, which, as he tells me, was not a quantum theory, but was supposed to yield the quantum as a result. This, Hans deadpans, was the problem with this theory. We discuss the rise and fall of Einstein, from the wonder year 1905, when anything he touched would "turn to gold," over to 1917 the year of general relativity, up until 1926, when with Bose-Einstein statistics he still was in the thick of things, at age 47. That, however, was the end of it: one more influential paper much later with Podolsky and Rosen, then assistants like (Peter) Bergmann, (Valentin) Bargmann, and Strauch, the latter pretty much unknown today. But, as Hans insists, he was a great man when he was young, and a sweet man when he was old.

Getting up from the table Hans says he has to make a phone call to Seattle. Gerry, nosy as ever tries to find out who he wants to call. After some prodding, Hans reveals the name of the person: "Pete Rose, not the one from Baseball." As he goes to the phone, he closes the door behind him. Gerry, writing a letter to Peierls to inquire about his health, says:

"He's cooking up something again, he never is that secretive. I practically had to pull this name out of him!" I wish Gerry a good night but he's already engrossed in his letter again.

Friday, January 31st

Lunch is with all the group today as every Friday, minus Hans for the reason mentioned last Friday. It seems I never really had an interesting Friday lunch at the Athenaeum, compared to those with Hans and Gerry. After lunch Steve Koonin comes into my office and says he wants to talk to me "about next year." We settle on Monday 2 p.m. Gerry already divulged to me that he's going to offer me a job.

I do not remember how, but at dinner we're in the middle of talking about history, and school, again. It seems to be one of Hans's favorite subjects. I get to tell the tale of my final (oral) exam in history for the Baccalaureate at the European School in Brussels, one of my favorite stories. The rest of the conversation is centered around Germany. First Hans asks me if German students are politically interested. I tell him that they are very much so, that almost everybody takes a side. I realize now that this might be slightly exaggerated, that it probably just seemed so to me because my friends where very involved in politics. I say that I was less involved, but that I voted in the Student Parliament elections every year. Since I had told Hans which parties could regularly be found at campuses all over Germany, he naturally wanted to know who I voted for. From there on we very quickly went on to the German political parties, and the rise and fall of the Social Democratic one. We talk about Helmut Schmidt, and Hans explicitly inquires about Oskar Lafontaine. Later the integration of East Germans is our topic, one which Hans always seems to turn back to. Also we discuss the "Volksdeutsche" (ethnic Germans in Eastern Europe) and their plight. Hans asks if all these people living in the Banat (the German enclave in Romania where my father grew up) would like to return to Germany. I venture the guess that their living conditions are probably very bad, but that they consider the region their land, which they had cultivated, and paid for with their lives, and thus would be rather unwilling to leave.[8]

Hans mentions a visit to Italy, with his wife in 1951. He describes how everybody was very nice to them, but very poor. He also mentions visiting the best-preserved Doric temple in Agrigento. Gerry asks if he also visited the place where he got arrested. A very thinly disguised invitation to tell a story. We're over dessert already: Mrs. Smith Cherry Pie.

"San Leonardo," Hans says slightly contemplatively.

"But we weren't arrested, we were under house arrest. It was 1933, the fascists."

"A friend of mine and I were vacationing in Tirol, and we were invited by his mother, who lived across the border in Meran. We wanted to walk over the border, so we consulted this guide, the "Baedeker," which mentioned several passes that were open. We chose the "Timmel" pass, which we walked. It was an incredibly boring walk, until we came into Italy, when

---

[8]In fact, almost all Germans left before 1945.

it was quite nice. A little later we met some border guards, who informed us that we could not take this pass. I spoke only a little Italian, and the guard spoke no German, so he took us to the border station. That was approximately 6 km away, and we were two young, tall, guys and well trained, and the officer had quite some difficulties following us. Then we arrived in this little town, and they directed us to a room where a "Maresciallo" was waiting, who wasn't really a "Feldmarschall" but a Sergeant, but he had a beautiful uniform. At the beginning we had to stand, and it didn't seem as if he understood my Italian. Then he asked what our professions were. I answered "libro docente." "Bring chairs," the Maresciallo immediately ordered, and started to understand my Italian. We described our situation, and he insisted that it was forbidden to cross the "Timmel" pass. However, he said that he would call the border station at the "Brenner." Until then, he asked us to stay in town, and we took a hotel. The next morning we presented ourselves at the little border station. The Maresciallo had contacted the Brenner station and they had decided that we had to go there and have our passports stamped. The only bus from the little town to the Brenner, however, had already left in the morning. So my friend ordered a taxi from Meran."

"That is far, that must have been quite expensive," Gerry asks.

"Well," Hans responds, "he was already quite well to do then."

Upon Gerry's further inquiry Hans reveals that his friend worked at a bank.

"Earlier the Maresciallo had told us that two soldiers would accompany us. When I told him that a taxi was coming to pick us up he exclaimed: 'In that case I will come myself!'"

"When we arrived at the Brenner station they locked us in a little room, then asked us to come out. They told us gravely that it was an offense to cross the Timmel pass, but that they would not fine us for that, and let us go free. We were walking up and down the platform waiting for the Maresciallo, but he was busy drinking wine with his colleagues! Finally we left the Brenner, dropped the drunken Maresciallo at his little village, and drove on to Meran."

Saturday, February 1st

Got up at 7 a.m. this morning. At 8 a.m. I had breakfast with Hans and Gerry, as we planned to go up to "White City," a hiking destination at the foot of the mountains here. This is an interesting hike that Gerry and I had already done two weeks ago. In fact, I found it quite strenuous then, and I was wondering how Hans would take it (so was Gerry, as he told me later). "White City" is the upper station of a trolley that used to go up the mountain, from its foot at Lake Avenue. It ceased to operate in 1938, and today "White City" is mainly ruins. Yet, one has a beautiful view down to Pasadena, the Jet Propulsion Lab, and downtown L.A. further back. The trail is rather narrow, such that we had to walk one behind the other. This unfortunately prevented us from communicating much, also Hans was concentrating on the sometimes treacherous trail. As he suffers from "Menier's syndrome," his sense of balance is somewhat reduced. Yet, he managed the ascent and the subsequent descent admirably, and this convinced him that he'll live to be 90, and see the results of the SNO neutrino experiment, after all. At the top, we ate our sandwiches quietly, while Gerry took a little nap. I was lying on my back contemplating the sky, which was partly cloudy and provided for partial shade during the ascent.

On the way down we stopped and had some water seated under the pylons of high-voltage power lines. The silly debate going on in the general public over the effects of electromagnetic fields on people prompted Hans to tell me a story, while Gerry was napping again. There was a public meeting of the city council of the city of Santa Barbara not long ago, for reasons of introducing the new candidates for the upcoming election for city council members. One of the candidates was the wife of Jim Langer, who is the director of the Institute of Theoretical Physics of the University of Santa Barbara, and a friend of both Hans's and Gerry's. He is in fact a former student of Gerry's, and I met him when I was down there last year to give a talk. The frightening moment occurred when someone from the audience asked the candidates if they would oppose or agree to an ordinance banning electromagnetic radiation. This is not a joke, as this question was seriously posed, and in fact the first three candidates all agreed that, yes, they would support such a city ordinance, to ban this terrible electromagnetic radiation. After this display of expertise Langer took the microphone and declared that although he was not a candidate, his wife was, and he wanted to point out, in his qualities as director of the Institute of Theoretical Physics, that they would have to ban the sun if they wanted to follow through on such an ordinance. The city council kindly invited Langer to give them a briefing when the vote on the ordinance would come up. Scary stuff.

We made it down the mountain at 3 p.m. We had a short discussion about the legalization of drugs, and whether such an act would better or worsen the situation. We were all pretty much undecided; however, we agreed that a legal distribution net could take away much of the violence associated with the drug war. As the drugs would be used whether legal or not, this would seem like an improvement. After all, alcohol is legal too.

Dinner at 7 p.m. Gerry asks Hans about the Emperor Franz Josef, as Hans had given him Gibbon's "Rise and Fall of the Roman Empire" to read. Interestingly, this sparks the only physics anecdote of the evening. Hans tells us of the fun that Viktor (Vikki) Weisskopf and Enrico Fermi had comparing their "impressions" of Franz Josef. To Weisskopf, Franz Josef was "der Kaiser, der sich um das Wohl seiner Völker kümmert." For Fermi, he was "il odioso Francesco Guiseppe." Or so they were told at school. Having branched into history, however, Hans is unstoppable. He goes back to Maria Theresa, going through all her children, with a detour over the Kings and Emperors of France after Louis XIV, the revolutions of 1848, to the assassinations of the heirs to the Austrian-Hungarian throne, the last one in Sarajevo, 1914. He knows all the details: Franz Josef born in 1830, enthroned in 1848, emperor for 68 years (or so, I don't remember as well as Hans). This is how he claims he knew Franz Josef's year of birth:

"I know that from a stamp that commemorates his birthday."

"You collect stamps?" I ask.

"Yes!" he answers.

"Me too!" I say.

"Very good!" he finishes. A very typical exchange.

It is quite amazing. He knows the kings of Venice, the one put in by Napoleon in Sweden (Maréchal Bernadotte, if you must ask) in fact he lists all those that Napoleon put in as Kings, the one the Austrians put in in Mexico, etc. etc. I cannot possibly list here all the things Hans told us this evening. (Also, since these are generally known, this would be quite repetitive.)

There is a brief discussion about aid to Russia during dessert. Hans mentions a group called "Direct Relief" that he seems interested in (in any case he has talked to one of the organizers of this relief effort). This group currently ships medical supplies into Russia. The way I see it, though Hans didn't imply it at all, he is, or will be, helping them financially. We talk about the ruble again. He says that there is no way of rescuing the ruble. They need a second currency. (Hence his letter to Yeltsin.) Hans speaks favorably about Yeltsin (just having read one of his recent speeches). He speaks favorably of Bush only with respect to arms control (and, after Gerry's inquiry, his behavior in the Gulf War), but he despises his domestic policy, as "doing more harm than good."

Sunday, February 2nd

We have lunch at 12:30, and are planning to go to the Huntington Library and Gardens later. Hans received a letter asking him to write an article for "Global Viewpoint," a group that has an agreement with most of the major newspapers in the world to have their articles published. It says that he would have a readership of about thirty million people. They then go on to enumerate some of the recent contributors. Every major head of state seems to be there. The topic they would like Hans to comment on is the so-called "Baruch plan" of 1946, which, it seems, is in the process of being revived. The letter states that among others, Edward Teller, and a Russian whose name I don't remember, have already embraced the plan. This is startling to Hans. Indeed, the Baruch plan was a singular attempt at limiting the use of nuclear weapons, by giving the monopoly of possession to the United Nations. According to this plan, no nation was allowed to build, or possess, nuclear weapons, and every nation found to do so would be punished, "severely and continely." Hans mentions that this was the first and last time he ever saw the word 'continely' used, and so have I, and Gerry. We decide to look it up sometime.

The plan originated not from Baruch, but in fact from Robert Oppenheimer. It was adopted by then-secretary of State Dean Acheson. However, it seemed to Acheson that this would never pass the Senate like this, and Hans adds that "he was most probably right." So the idea came up to convince Bernard Baruch to present it to the Senate, which was a clever idea, as there probably was nobody more conservative than Baruch. Also, he was hard to argue with since, as Hans says, "he was even more hard of hearing than I," and he would, during an unpleasant discussion, just turn off his hearing aid. Somehow they got Baruch to present the plan to the Senate, and so it bears his name. This plan now seems to have been revived, and it also seems to make a lot of sense, the only problem being China, who probably would refuse to comply. Hans says that he will agree to write the article, but that he will first request from "Global Viewpoint" an article or some information showing that Edward Teller advocates this plan. This indeed sounds incredible as Teller is (or was) one of the extreme hard liners, a hawk, anti-communist, nuclear supremacy advocate, and one of the staunchest proponents of the SDI. He and Hans are not talking, when politics is concerned.

We leave for the Huntington Library later, which is a complex featuring a Library with old manuscripts, an Art Museum, and a beautiful botanical garden and park, just a couple of minutes away. While walking in the cactus garden we encounter Stephen Hawking (who visits Caltech as regularly as does Hans), riding in his electrical wheelchair, somehow slumped into one corner of it, head crooked to the left, an electronic communication device with a screen in front of his eyes. He was accompanied by a woman. Even Hans and Gerry didn't dare to talk to him. Hans enjoyed the garden immensely, in fact he's been there several times and knows what is where, and studies some plants carefully.

At dinner Hans mentions that he talked to Rose, his wife, and that she was pleased about his getting involved with the Baruch plan. Gerry mentions that she was never pleased by Hans's participation in the hydrogen-bomb effort. Hans mumbles something, then says that he has "atoned since then." But naturally, the subject is now the H-bomb. I feel a certain reticence on Hans's side to talk about it (which I may have been imagining) and so I ask about the Soviet effort instead. I ask where they at all got the idea that a hydrogen bomb would work.

"This," Hans says gravely, "Fuchs told them. He didn't know any details then, just that it would work."

"He may have told them about the lithium hydride," Gerry asks. Hans answers somewhat distanced:

"This I don't know."

But he goes on to talk about Sakharov. He says that Sakharov was their mastermind, and that he presented the group with three designs, the nature of the second one he never revealed. The first one already had lithium hydride, and the third one worked. They had a preliminary test, not of the bomb itself, in 1953, and a successful test of the bomb, if I remember correctly, in 1955. Hans credits Sakharov for revealing that the first successful test came that late, as American intelligence claimed that the 1953 test had been successful in order to speed up the American armament effort. Hans had predicted earlier that it would take the Soviets three years to build the bomb, after the first successful American test showed the Soviets that it could be done, and he was right. Hans also recounts a story told by Sakharov, that at the dinner celebrating the successful Soviet H-bomb test, Sakharov gave a speech saying that they should be happy that they accomplished the goal, but that they should hope that this weapon would never be used. A Soviet general rebuked him. Giving a speech himself, the general said that, on the contrary, this weapon should be used, to wipe out the capitalists. Gerry asks whether the Soviets were capable at this time of launching such a bomb on an intercontinental missile. "Indeed they were," Hans responds, "they were ahead of the Americans in rocket technology."

He then recounts, very calmly, the events that took place at a particular test of such a rocket. The military staff was present, including the general who had advocated the use of the bomb, and as the countdown receded to ignition, nothing happened. "Like the engineer in this story," Hans reports slightly smirking (I will tell the story he alludes to later), "this general goes up to the rocket to check the exhaust. At this point, the rocket ignites. He was fried to a crisp!"

"That was like a sign from god!" Gerry jokes.

"Yes!" Hans answers, now beaming.

We go on to discuss Soviet and American rocket technology. This being more my domain, I do most of the talking. We compare Sergei Korolev to Wernher von Braun, and the "American Germans" to the "Soviet Germans." Hans points out the accomplishments of the American team, especially the Minuteman, I point out that the Soviets would very probably have been the first on the Moon, if Korolev hadn't died so soon (I hazard a guess at 1962–1963).[t] After that, I point out, they basically tried to scale up Korolev's design, up to the gigantic N-1, which never flew, but exploded in three successive tests. America, on the other hand, had von Braun, who was very much alive and turning on the heat down in Alabama. Hans likes that.

Hans and Gerry relate their experience with the change in attitude that was being felt all through America in the late fifties, initiated by the "Sputnik Shock" and Eisenhower's reaction to it. Suddenly, everyone wanted to hire scientists. Gerry tells how suddenly it was no problem for him to get a passport to accept an invitation by Niels Bohr to go to Copenhagen, and Hans relates how he got an offer by the Chrysler Corporation to head their rocket development division, which he turned down. Instead, he says, he worked for "Avco," which seems to have been a conglomerate of companies, the exact nature of which Hans didn't reveal.[u] He does reveal, however, that he was working for them on "the re-entry of war heads." He says that the atmosphere was very pleasant at Avco, as long as the old chairman, Arthur Kantrowitz, was there. After he was replaced, Hans didn't like it there anymore and quit. Incidentally, I just remember that when Hans mentioned the mysterious Pete Rose that he had to call, he answered, upon Gerry's insisting prying, that Rose was "an applied physicist that he met at Avco," a name that didn't mean anything to me then.

From discussing the importance of the "Sputnik shock," which led Eisenhower to emphasize science, research, technology and engineering, we

---

[t] 14 January 1966.
[u] Avco Everett Research Laboratory, in Everett, Massachusetts.

draw analogies to the situation of today, the "Japan-shock," (or should we call it the "Toyota-shock"?) and discuss methods of dealing with it. Hans is convinced that the American people can stand up to the challenge, if they are made aware of the dangers, and if a program is set in place. I ask Hans if he thinks that Bush is aware of the situation, and of the steps that have to be taken.

Later, during dessert, we discuss a little bit of physics, something that Gerry brought up during lunch, and that we had figured out in the meantime. He tells Hans that there were "many ways to show this," but that he'll show one of them to Hans the next day. This sparks another story, and Gerry initiates it by saying that this was precisely what Ernest Lawrence had told Hans in Berkeley, when Hans was inquiring about the solution to a certain problem. (This was in 1936–1937, as Hans mentions upon my inquiry.) Hans interrupts, to tell the story of how it really was. Hans had written a paper[v] with a postdoc of his on the problems encountered in the physics of synchrotrons, namely that in order to focus the beam, the field strength had to get smaller towards the edges, whereas in order to counteract the effects of relativity, the field had to grow from the inside to the outside. From this he derived a "maximum energy for synchrotrons," which turned out to be 20 MeV.

"Ha," exclaims Gerry, "another mistake. First he claims we would see all those neutrinos from the Sun, and then he derives a maximum synchrotron energy!" Hans continues undisturbed on how he got angry letters by Lawrence in Berkeley, claiming that this wasn't right. He suggested that Hans should talk to Robert Wilson, who then was but a graduate student. (Later, he would head the "nuclear experiment" division at Los Alamos.) As this didn't lead to an answer (namely the bound would still be firm) Hans talked to Lawrence himself. So Hans asked him how he intended to show that there was no such bound on the energy.

"There are many ways to skin this cat!" Lawrence exclaimed, but Hans asked him to show him only one of them. In the end he failed. The problem was solved eight years later by McMillan, Lawrence's assistant, who showed how you can use amplitude modulation to circumvent the bound. He got the Nobel prize for this clever trick, and today's synchrotrons are still using it. Hans mentions briefly that "the people at the Berkeley lab" were quite reactionary, and hard to deal with, especially Lawrence and Alvarez. Now this is not so anymore, however Lawrence Berkeley Laboratory is not one of the world's top labs anymore.

---

[v]H. A. Bethe and M. E. Rose, *Phys. Rev.* **52**, 1254 (1937).

The story that Hans alluded to when he talked about the general who was "fried to a crisp" was told yesterday, when we were in the middle of discussing Maria Theresa, and her children, and the French Revolution. Maria Theresa had a lot of kids, many of them girls (Hans would know the exact number, and the names). All the girls were called Maria-something. One of the "somethings" was "Antoinette." "She came to a sticky end," Gerry quipped, and then proceeded to tell the story of the two noble men and the engineer, who are scheduled for execution by the guillotine. The first noble man goes and puts his head under. The blade falls, but a snag holds it up just before cutting the head. According to the prevalent law, the noble man is released. The second noble man is forced on the instrument, and the same thing happens, upon which he also goes free. Now it is the engineer's turn, but instead of presenting his neck to the blade, he presents his throat. Looking upwards, he points at the instrument and exclaims: "Ah, there is the problem!"

Third Week

Monday, February 3rd

Hans finally got Yeltsin's letter. He wanted to show it to me, but didn't get around to yet. From what he told me and Gerry (who already saw the letter) Yeltsin seems to consider Hans's proposal. At lunch, more anecdotes about Mrs. Peierls. I suppose that there is an endless supply of them. Once, Hans let his son Henry, then still very young, in the care of Mrs. Peierls for two weeks. A couple of days after he picked him up, Mrs. Peierls comes to dinner to the Bethes. They open the door, and this was when Henry Bethe spoke the first word he ever pronounced. He saw Mrs. Peierls and went: "Nooo!"

For part of the lunch Gerry and I talked about our longer manuscript.[w] He's afraid that I won't finish it, I'm afraid that he sends it off as it is. I promised him that I'll get the job done, and he promised me that he won't send it off without my permission. "Of course," he adds, "that doesn't mean that you will be sober when you give me the permission. I have a lovely bottle of Chianti at home; it'll make you sign anything!"

After dinner Steve Koonin talked to me in his office and offered me the Prize fellowship. I accepted. Dinner is at 7 p.m. Gerry made two chickens. I tell Gerry and Hans how I am improving my record over two miles of running

---

[w] *Phys. Reports* **234**, 1 (1993).

(currently at 18:04, I'll break 18:00 tomorrow, my goal is 15:00), and Hans tells us how Freeman Dyson's wife has taken up running the marathon, and is actually performing well in her age group (of fifty to fifty-nine, she being at the lower end.) This is how it goes at these dinners: free association.

Hans asks Gerry what made him join the communist party at Yale. Gerry complains that Hans always asks these difficult questions. He first goes into explaining how, when he was in the Navy, he shared everything with everybody there: Blacks, Jews etc., but when he came to New Haven he found instead ghettos and discrimination. Somehow, though, I get the feeling that that wasn't all that did it, but he doesn't tell more. He goes on talking about the "Wallace movement," and about the FBI's infiltration of that movement, of tapped phones and people following him. Finally, he and a good friend (and physicist) of his are expelled from the Party for alleged "left-wing deviationism." Gerry leaves the country early enough towards Birmingham, but his passport is revoked. We talk about the communist witch-hunt going on at that time at other universities. Gerry mentions Harvard, and how the faculty there stuck to Prof. Furry, who had a communist past. Indeed, in one year all faculty donated 2% of their salary to pay the legal fees incurred by Furry. Another time, Furry was supposed to talk on television. Because they were afraid that he would expose himself, they asked the physicist Norman Ramsey to talk instead. So he did, and sure enough McCarthy was watching the show. Afterwards, McCarthy invited Ramsey to dinner, and asked him how much money he was making at the university. After Ramsey told him, McCarthy answered: "I'll triple your salary if you come to work for me as my P.R. man!" Needless to say, Ramsey refused.

At dessert (we have pie) I ask Hans whether he knows Ramanujan's representation of π. He says that he didn't know that one, but that he knows a story about Ramanujan. Indeed, Ramanujan, who was a self-taught mathematician who "dreamed up" formulae without proof, which often enough, however, turned out to be correct, was known to find a number-theoretic peculiarity with every number. Once, the mathematician G. H. Hardy (who brought Ramanujan to England) visited him while he was recovering in a London hospital. He mentioned to him that he had come over in a taxi, with the number 1729. "Surely, you will find nothing peculiar about the number 1729!"

"But on the contrary," Ramanujan answered immediately, "this is a very interesting number. It happens to be the first number that can be represented by two different sums of cubes!" Naturally I asked Hans if he knew what these two sets of numbers were. "Of course I know!" he answered.

"You just have to know your cubes." How could I have doubted. Unfortunately I couldn't find it fast enough, so he told me. *You* will have to figure it out for yourself, though!

As we were talking about numbers, and tricks with numbers, we automatically came to talk about Feynman. According to Hans, Feynman was a wizard with numbers, by anybody's standards. He tells the story of one of Feynman's semi-official visits to Brazil, when he went into a bar where a numbers artist challenged people out of the public. It was made for Feynman. He won hands down. As the last challenge, the number artist said: "Cube roots!" And he proposed the number ... 1729! That of course was easy money for Feynman. They both sat down to calculate, but Feynman had the first three digits after the 12 in a matter of seconds. While the magician was checking this answer, Feynman found three more ...

Gerry asks Hans if Feynman was happy. "Very," Hans answers, "most of the time." We talk about Feynman's wives; his first wife died in Albuquerque of tuberculosis while he was working on the bomb in Los Alamos. She was extremely bored in the hospital, so they wrote each other many letters. At one time, they decided to write each other in code. This proved to be very unsettling to the authorities, as all letters going out of Los Alamos were routinely censored, and the letters thus kept coming back to Feynman. He finally decided he would include the key to the code with the letter, so that the censors could read it too. Still, this proved to be too much for them. The second wife, it seems, pursued Feynman for three years until he finally "gave in." Hans says that the woman wasn't much good. The third one, a British woman, however, everybody seems to hold in high esteem.

Another story: at Los Alamos, there were of course very strict security measures. However, Feynman discovered a hole in a fence surrounding a toolshed, from which he could leave the center without checking out. Three times in a row he did just that, entering without having checked out, until a confused guard finally questioned him about it. I suppose Feynman loved to confuse people. Hans also tells the story about Feynman cracking the five safes of de Hoffmann, a story that appears in his "Surely You're Joking ..." book. He tells it very comically, but since it already appears in print, I won't repeat it, as I didn't discover anything new in the story. I don't remember if de Hoffmann's name was mentioned in Feynman's book though. Hans says that de Hoffmann died several years ago of AIDS, acquired through a blood transfusion. I ask Hans if what Feynman did was real or if he was just using tricks. "Everybody uses tricks!" Hans replies, this of course not being an answer. Gerry throws in that Hans isn't above using tricks himself.

He maintains that Hans's motto is: "Never enter a competition without an unfair advantage." Hans grins broadly. I am very sure that Hans is great with numbers too, but I'll ask for a demonstration another time.

At lunch, Gerry or Hans told a story where Hans was invited to a dinner by a colleague. After being introduced, the wife of the colleague asks Hans: "Oh, so you are the one who writes the *Physical Review*?" Says Hans remembering that: "These were the olden days!"

Tuesday, February 4th

Tuesdays at noon is the Journal Club at Kellogg, and Gerry made sandwiches for Hans and me. He spoke for half an hour on deep inelastic scattering; a graduate student had the other half on J/ψ.

At dinner Hans asks me what was going on in the J/ψ talk. As the girl spoke very softly, I suppose he didn't catch all of it. I explain to him the main points, and he is satisfied. Funny enough, we spend most of the dinner reciting German children's poems. Hans knows a lot of them by heart, so does Gerry. In fact it turns out that I am the one who knows the least! We discuss the tales of Grimm, and recite Heinrich Hoffmann's "Max und Moritz," and the often frightening tales of "Der Struwwelpeter." It is unique to hear Hans recite:

> Paulinchen war allein zu Haus,
> Die Eltern waren beide aus.
> (Mamma and Nurse went out one day,
> And left Pauline alone at play.)

or

> Und Minz und Maunz, die Katzen,
> Erheben ihre Tatzen,
> Sie drohen mit den Pfoten,
> "Die Mutter hat's verboten!"
> (When Minz and Maunz, the pussy-cats, heard this
> They held up their paws and began to hiss.
> "Meow!!" they said, "me-ow, me-ow!
> You'll burn to death, if you do so,
> Your parents have forbidden you, you know.")

etc.

Hans mentions an article he read in the newspaper about an archaeological discovery, namely that they found traces of *Homo erectus* in Tbilisi (in Georgia) that are 1.7 million years old. This, he says, is quite astonishing as *Homo erectus* supposedly originated in Africa, about two million years ago. That the species travelled so far was up to now unknown. Since neither I or Gerry are experts in this kind of history, we let Hans tell us a bit about it, and so he does, about which species was where, the relation to other species, he gives us a time table of events; as always he is a walking encyclopedia. We venture into theories of the extinction of the dinosaurs, and Hans offers his explanation of the extinction of the saber-tooth tiger, which he says was a contemporary of early Man. He says that they were such excellent hunters that they basically killed off the species, which were their main food. I ask him if this is established wisdom, and he says yes. I find that strange as I would expect that the species would, in such a case, survive by feeding on themselves, which apparently however they did not do. Strange story.

This leads us to the problems of our world, which potentially could lead to the extinction of the species "Man," which in my view is inevitable should that species turn out to be "a bad idea," as all of nature's bad ideas disappear sooner or later. Hans doesn't think that mankind is a case of a "bad idea," however we discuss the trends of growing world population, and lessening rain forest area. We try to figure out how much area of rainforest is needed to offset the CO_2 emission of a car engine, but fail, not having enough data. Hans believes that both problems (of world population and rain forest) are solvable. I believe that the rain forest problem is solvable in the near term, and that the other problem will linger until very close to disaster.

Hans still wants to get some packing done, so he retires (Gerry is still busy doing Hans's laundry). I wish Hans a good trip, but he says that we'll still have lunch together tomorrow before he leaves. So be it.

Wednesday, February 5th

Last lunch with Hans before he leaves for ten days. He'll be back for five days after that. Not a very chatty lunch. I guess that Hans is in "travel mood." Gerry mentions that Felix Boehm of Caltech will have open heart surgery the day after tomorrow. In such circumstances Hans says: "Gosh!" After weighing the pros and cons of exercise, and whether to choose surgery or not, Hans says: "Sometimes it's a pity that I don't believe in God, because now I don't have anybody to thank for that I don't have to worry about such things!" The rest of the lunch is physics, interrupted by periods of silence. Hans wants to know why chiral symmetry is broken on the lattice. Our answer is of the sort: "What do you mean, why?"

Hans wants to get going quickly, no time for any farewell. I already wished him a good trip yesterday. He just shrugs off any attempts: "Ah, I'll be back in ten days."

I have dinner with Gerry alone at 6:30. We are eating spaghetti and leftovers from last night's dinner. I mentioned to Gerry earlier that day that Steve Koonin had told me that I was awarded the Prize Fellowship, which I had accepted, and for celebration he opened the bottle of Chianti that the Goodsteins had brought for dinner, a few days ago. It turned out to be excellent. Gerry talked mostly about my future, and related things. He told one story about Gregory Breit, his advisor at Yale. When he was in Minnesota, Eugene Wigner joined him there, and they decided that they should form a Nuclear Physics Institute. As they were at this point the only members, Breit proposed that they should choose a president. Wigner argued that they wouldn't need a president.

"O.K.," said Breit, "so *you* be the president!" Wigner reiterated that they didn't need a president.

"All right," said Breit, "then I'll be the president!" and as he was saying it punched Wigner in the face such that he collapsed on the floor. I couldn't believe this when I heard it, but recalled some of Gerry's earlier stories where Breit demonstrated physical violence against his students. Oddly enough, they wrote their most famous paper together after that incident. This story was told to Gerry by Wigner himself after the latter was somewhat tipsy following the Ph.D. celebration of one of his (last) students.

After dinner, Gerry and I went to South Pasadena to watch the movie "High Heels," by Sergio Almodovar at the "Rialto."

Saturday, February 15th

Gerry and Hans came back in the afternoon. Gerry surprised me in the office around 6 p.m., and collected me for dinner. We also called Tetsuo Hatsuda who was visiting for a seminar, from Seattle, and went to a little Thai restaurant on Colorado Avenue where Gerry and I were greeted with "How are you?" and "How is your son?" Gerry and I had been there before, and so had Gerry and Titus.

Hans was in a good mood, which may have been attributable to his getting something to eat (after the dismal lunch on the plane, as I was told). As is perhaps inevitable, I soon was in the middle of telling stories from my trip to Thailand in 1982, and we never really got away from that subject, as Hans wanted to know more and more about the Akha hill tribe. As I had spent a couple of nights then with the tribe, he didn't let up. He also was emphasizing how dangerous the North of Thailand is, something that I wasn't quite aware of at the time. When I mention the kind of stilted huts the Akha (and their visitors) slept in, Hans is reminded of very similar huts that he saw when he visited Cambodia, in 1969. This was before the Khmer Rouge took over, and Hans muses at how peaceful the country was then, and how unthinkable the atrocities committed by the Pol Pot regime were, murdering about one-sixth of the population. We agree that it probably was one of the worst genocides in the history of mankind. We talk about Angkor Wat, the 1100 AD Hindu temple that sits on the border between Cambodia and Thailand, and Hans bets that it should be possible to visit it in about two years.

Tetsuo didn't say much all evening, he probably was feeling much like I felt when I first met Hans. After dinner Gerry asks Hans if he is sated, and Hans answers "Mäh, Mäh!" alluding to the line "Ich bin so satt, ich mag kein Blatt, Mäh, Mäh," from a fairy tale of the Brothers Grimm that Gerry and Hans are quoting off and on.

Sunday, February 16th

So many stories, so little time! Tonight we had dinner with Vendula and Petr Vogel. She is a (math) teacher and Academic Dean at Westridge School for Girls in Pasadena, knows a lot of the names and stories, and the latter abounded. I can only try to pick out the cherries. He is a nuclear physicist at the Kellogg Lab, and they seem to know Gerry from far back, easily twenty years. But first, our morning hike.

At breakfast, Gerry surprised me by asking me what the r-process was. He flatly admitted that he only learned about it at the conference that he was at with Hans just now. I, however, must concede with no trace of humility whatsoever, that I knew about the r-process, and I will add with even more candor that I only learned about it such that, in the event someone would ask me about it out of the blue, that I could just answer off-hand. Indeed, I ran across a *Physics Reports* article[x] about it a few months ago when I was at the University of Regensburg, and read it. It is nuclear astrophysics, so it was completely irrelevant for my work, but it was kind of interesting. And secretly, I hoped that someone would one day try to "catch" me ignorant of it. This sounds like a silly game, but if you do this often enough, it pays. I

---

[x] "The r-process and nucleochronology," *Phys. Reports* **208**, 267 (1991).

sort of looked around and said: "Well, yes," and explained. Gerry tried to catch me again with the s-process:

"What does the 's' stand for?"

"Slow," I answered calmly. Gerry was mightily impressed.

"Why is it slow?" He wanted details. I remembered them while I was speaking. Later in the evening the same topic came up again, because Gerry likes to joke about my lack of knowledge about nuclear physics, telling Petr Vogel he should teach me some when I'm at Caltech in the fall. To which I just remarked that at least I knew about the r-process! Gerry conceded, and I told the story of how I read about the r-process while in Regensburg, just because the article was there (which was true), and Hans liked that a lot, agreed that "this is the right way." The moral is: sometime these little games I play pay off with dividend!

Last week we had the worst storm in Pasadena for the last fifty years, so the media said. Indeed, and I was right in the middle of it. It did not stop raining for a considerable amount of time (like, four days), and the California highways are just not built for this kind of beating, to which they respond with flooding. This is, however, only a temporary impairment, but when we went to the foot of the mountains to walk a specific trail this morning, we got a glimpse at the expanse of the damage. The entrance to the trail was blocked by a fence; in the valley beneath us we could see and hear a white river in a bed that probably sees a few drops of water only once a year, and the trail, well we could see the rest of it on the side of the mountain, it was simply washed away in gigantic mud slides that came down from the mountain. No hiking on that trail. So we decided to drive to another area where we could walk on an asphalted road. During this brief trip, we got to talk about David Hilbert, professor of mathematics at the University of Göttingen for a long time. Sure enough, Hans had some stories about him. Indeed, the way we broached this subject was on the heels of a discussion where we realized that at each phase of our career, while being a student or later, there would always be someone who was just better, and smarter. But also how these people somehow got left behind at the next phase. Gerry claims that for him it was because he worked hard. Hans countered that he also knew people who were smart, and worked hard, and still they never amounted to anything. He suspected that those just simply lacked imagination. This is where he quoted Hilbert. The way he rendered the quote, it was clear that every word was authentic Hilbert, transmitted to him, as I learned later, by his father-in-law Paul Ewald, who happened to be "physico-technical assistant" to Hilbert at one point in his career. Hilbert was talking about an ex-student of his, and the exact quote is this:

"Also der Schmidt, für die Mathematik hat er nicht genug Phantasie gehabt. Jetzt ist er Dichter geworden. Dafür hat es gerade gereicht."[y]

He related to us another quote of Hilbert during the short ride, and that was in the middle of the 1930's, when the mathematics department had been all but completely wiped clean of the Jewish faculty, and replaced by Nazi followers. Hilbert, not being Jewish and therefore untouchable, was asked by a Nazi if it was true that the level of mathematics in Göttingen had gone down. "Das würde ich so nicht sagen," answered Hilbert (Hans was quoting Hilbert in German), "die Mathematik ist nicht heruntergekommen hier, sie existiert einfach nicht mehr!"[z]

There was no interesting discussion during the hike, which took only about two and a half hours (it started to drizzle slightly). I remember some talk about physics, and attempts to locate downtown L.A. from our vantage point. Indeed, due to the previous rain the atmosphere was pretty clean despite the low-hanging clouds, and L.A. appeared to be very close. Hans and Gerry claimed that the structures that we could see in the distance were Glendale, and I, because I had memorized the vista on earlier hikes, maintained that it was L.A. After the first discussion I actually gave in to the combined pressure of Hans and Gerry, after I asked Hans whether he was sure that this was Glendale. This might seem like a trivial point all in all, but then the reader doesn't realize how points like this are of tremendous importance to at least the three of us. Everybody wants to be right, nobody wants to be wrong, no matter what the object of discussion. The discussion about this never got heated. I gave in after Hans and Gerry both claimed there were sure. On the way back it was clear they were wrong. I first convinced Gerry, and he happily swung to my side, as he could show how Hans was wrong again. Grudgingly, Hans admitted defeat. We may have very erudite discussions most of the time, but we behave like children if it is about who's right and who's wrong.

On the way home in the car, suddenly Hans says: "Good!" I am somewhat startled, and Gerry immediately asks what was the matter.

"I looked out the window to the names of the cross-streets," Hans replies.

"I saw one which said 'Atchison,' and I wondered if the next one wouldn't be called 'Topeka.' And indeed, it was." Gerry smiles, I'm completely at a loss. How do you deduce that the name of the next cross-street will be

---

[y] "So this Schmidt fellow, in the end he did not have enough imagination for mathematics. Now he is a poet. For that he had just enough."
[z] "No, I wouldn't say that mathematics has declined here. It simply doesn't exist anymore."

"Topeka," if you just read "Atchison," not being familiar with the neighbourhood of course? Well, the answer is in one of the first transcontinental railway lines, which was the Atchison-Topeka-Santa Fe Line, as they lectured to me subsequently.

"The interesting thing about this line is that it never connected to Santa Fe!" Hans adds. I suppose that, if you have lived that long, and never forgot a single fact, you come across information like that.

For dinner, Gerry had made his goose, once more. For hors d'oeuvres we sat around the couch table with a Californian Chardonnay, that I bought ("Get a light fruity white wine, I don't know how to shop for wine," Gerry had asked me). Somehow Mrs. Vogel seemed to praise the American husbands over the European ones (the Vogels are of Czechoslovakian origin) which I, maybe slightly rudely, contradicted. ("This is absolutely wrong.") I realized that she was maybe a bit disconcerted by that, so I didn't say anything for some time, but rather listened, to let her recover, then made polite comments about her tales about the new developments in Prague, which they visited recently (they still have many relatives there). Hans was very interested in stories about Prague and Czechoslovakia, and bombarded Petr Vogel with questions. I was "sommelier" like always, so I didn't get much out of it. At dinner, the conversation turned to gossip about famous and not so famous physicists that had crossed the paths of those gathered. It is completely impossible to remember more than a few. The circumstances of Richard Feynman's death were elucidated (the Vogels were friends with the Feynmans) and Mrs. Vogel told how Gwyneth Feynman didn't invite anyone from Caltech to the official viewing of the film that the PBS station had made about Feynman, and his death. The reasons seemed to be that Gwyneth Feynman thought that Caltech didn't really acknowledge the eminent stature of her husband, and also that she just didn't get along with the other faculty wives. Then Murray Gell-Mann was the object of our attention. "He seemed to continuously struggle against the overpowering image of Richard," Mrs. Vogel claimed. "He liked to maintain that the V-A theory was really his idea" at which we all just snicker. Gell-Mann of course is the one who invented quarks, and a large part of QCD. I relate the story of how Gell-Mann claimed that he gave Feynman a problem to work on, after the latter had asked him for one. "Quantize gravity!" he is supposed to have said. After running into serious difficulties, Feynman goes back to Gell-Mann and tells him that it's not working. "So, I propose something easier. Quantize Yang-Mills!" Feynman encountered basically the same difficulties, and decided they were generic. Indeed, he had discovered the so-called "ghosts."

He never wrote it up, though. The only record is the famous *Acta Physica Polonica* transcript of a talk he gave. I already mentioned this article earlier. I was told this story by Max Dresden; he insisted that this was Gell-Mann's view of history. Mrs. Vogel tells of other troubles of Gell-Mann. He seems to be an art collector, and once bought illegally exported art objects of Peru, for over $100,000. The police found out, and he was forced to return the objects, without refund. He avoided being criminally charged, and he even avoided to be mentioned in the L.A. newspapers. I asked who owned the LA Times. "The Chandlers" I was told. "They were good friends of the Gell-Manns. They live just a couple of blocks from here."

A 1964 conference is mentioned that both Hans and Petr Vogel attended (without them knowing each other). Petr Vogel claims that Feynman didn't go because he didn't want to shake hands with Pontecorvo. Hans doesn't believe that, that this would be very unlike Feynman. Vogel hints at a dark past of Pontecorvo, that he assumes is common knowledge (and probably is, but just not mine)[aa] and that he later was at the Rossendorf Institute in Dresden, with Klaus Fuchs. (I begin to see.) Hans says: "I did not mind Pontecorvo. I did mind Klaus Fuchs."

As we already went over Feynman's death, we go over some more physicist's demises. Petr Vogel mentions one physicist who died "like a man, skiing, in Dubna." There seemingly are very many opportunities for trail skiing, as Dubna is essentially flat. He died of a heart attack. I mention Heinz Pagels, who died while on a hike with other physicists. The Vogels know the story very well, as they arrived a few days after the incident. Somehow Pagels lost grip on the trail, or the trail crumbled under him, and he just fell, over 100 feet. When the rescue helicopter came it was already dark, and they had to wait until morning. By this time Heinz Pagels was dead. Another physicist's death is mentioned, this one while rock climbing. It is already very late while I write this, and I don't recall his name.

The Russian physicist Efimov is next on the list. I've known his name for a long time, as I took (Werner) Sandhas's course on scattering theory in Bonn and learned about few-body problems. There is a "Efimov effect." Efimov, the story goes, is in the United States trying to get a job. Vogel mentioned that he was at Caltech for six months, but that he didn't really do anything. The same seemed to have happened elsewhere. Gerry says that he didn't do much in the last fifteen years. Mrs. Vogel mentions that he last

---

[aa]Bruno Pontecorvo was an Italian nuclear physicist who fled to the Soviet Union in October 1950, 10 months after the arrest of Klaus Fuchs. He died in Dubna in 1993.

tried to get a job in a supermarket, but that he was turned down because of his accent. I can scarcely believe it.

Some more Hilbert stories. Hans says that some of them are written down, others were related to him by his father-in-law Ewald (as already mentioned), and some by Courant, who was Hilbert's collaborator. The Vogels know a mathematician by the name of Erica Toth, she worked with Hilbert. According to them, she dressed terribly. They claim that Hilbert once arrived at the Institute with an iron, such that Toth would iron her clothes.[bb] I ask if this was typical of Hilbert. Hans answers that the iron was most certainly an idea of Grete, Hilbert's wife, and that Hilbert was terribly absent-minded. Once, Hilbert was to have a discussion with a German cabinet minister. He was sitting in the waiting room, when he suddenly would burst out to the corridor, open the window, and call out to his wife: "Grete, what is it that I wanted to discuss with the minister?" Another time, Hilbert was giving a big party. When everybody was there, it turned out that only Hilbert himself was missing. After a brief search they found him asleep in his bed. He had started to change for the party; but after undressing obviously forgot about the party, and simply had gone to bed.

This astounding fact was revealed by Mrs. Vogel: Gerry had a native American grandmother, and that this is the reason he wasn't getting any grey hair! Gerry didn't care to comment, other than that this had already been revealed in a book that was published in France, and which told the biographies of eight famous men that the biographer had known. I will have to ask Gerry again what her name was.[cc] As the seven other people were all well-known Frenchmen, the question of the day at the time of publication was: "Mais qui est ce 'Gerry Brown'?"[dd]

Monday, February 17th

Back at lunch, 12 sharp at the Athenaeum. As reliably, within we are in the midst of Napoleon's offensive against Moscow. Hans presents us with the mysteries of the Fires of Moscow, more precisely the mystery of who laid them. Indeed, Napoleon was in Moscow, but the Russian Admiral Kuznetsov wouldn't give battle, but retreated. Napoleon had counted on

---

[bb]Note: I could not find any mention of a mathematician Erica Toth. The only female assistant to Hilbert I could find was Emmy Noether.
[cc]Dominique Saudinos, *Leurs leçons de vie* (Mercure de France, Gallimard, 1990).
[dd]But who is this 'Gerry Brown'?

getting supplies in the city, but found nothing. With the fires in the city and the citizens fighting in guerilla tactics, Napoleon had to retreat. This Hans compares to the retreat of Wellington into Portugal, and the unsuccessful retreat of the Austrians, as the Napoleonic supply lines were never broken. We also dwell upon Hitler's defeat so close to Moscow that he could see it. Hans suggests that "for your education" I read Shirer's "Rise and Fall of the Third Reich," which he found to be very accurate. Next he discusses the cutting-off of the British supply lines by the German submarines in 1942, and how the German submarine fleet was defeated. Indeed, due to the need of having a tube protruding over the water to supply oxygen, the submarines could be detected by radar. This was, as he says, later overcome with the invention of nuclear submarines. I ask Hans who invented those, and he tells how right after Fermi had shown how the nuclear reaction would work, Westinghouse and General Electric started working on a reactor. It turned out however that these reactors were much more expensive to build than the conventional ones. A navy captain finally had the idea to use them for submarines. He pushed the development, and became an admiral later. (Hans knew the name: Hyman Rickover.) For the rest of the lunch I'm afraid I was playing the entertainer, with stories about Apollo 13, other accidents in space, material left by the Apollo astronauts on the moon, footprints on the lunar surface, the Martian atmosphere, and my preferred subject: Titan, one of Saturn's moons. What can I say, I'm a space buff.

For dinner we have "The Remains of Goose" from the previous night. Hans begins to tell me about something that he just learned, namely that you cannot have turbulence in two-dimensional systems in the absence of viscosity, simply because of angular momentum conservation: smaller "rings" cannot be shed by larger ones because that would, at constant energy, raise the angular momentum. He is investigating the effects of convection in supernovae, and he was always told that convection would not lead to mixing (which he needs). However, these models of convection were two-dimensional, and thus without turbulence. A real three-dimensional convection system would exhibit turbulence, and lead to significant mixing.

After a short lull in conversation Hans announces that Governor Paul Tsongas of Massachusetts had won his heart again (as a Presidential contender), because he was quoted in a newspaper that he supported nuclear energy. We all agree though that all Tsongas can do is lead the way for Cuomo in 96, as no Democrat will win in 92. I ask Hans about Cuomo's stand on nuclear energy, and Hans responds: "Not good." Cuomo seems to be generally in favor, but vetoed the Shoreham plant after it was ready to go on line, a six-billion dollar instrument. I quote the concerns about emergency evacuation, and he responds that those were entirely manufactured. After all, he said, all you need is to be five miles away, and out of the prevailing wind direction. This cannot prove to be difficult.

While Gerry is on the phone several times, I ask Hans about his famous "linear chain" paper, which was published in 1931, as he remembers.

"It had its 60th birthday last year!" Indeed, it was unknown to him that the now so famous "Bethe ansatz" was taken from this paper. When Feynman told Hans shortly before his death that he was getting interested in the Bethe ansatz, Hans answered genuinely surprised: "What is that?" Hans wants me to tell him what this Bethe ansatz is all about, but I can only tell him that it is some sort of solution to some two-dimensional conformal field theories, I really know next to nothing about it. I tell him that I should maybe read the paper, as I never did. He answers that he hasn't looked at it for pretty much sixty years himself.

I ask him about number games, and he says that in order to sleep he likes to take a number and find out whether it can be decomposed into a sum of two squares. For instance, he goes on, there is a mathematical theorem that states that every prime number of the sort $4n+1$ can be decomposed in such a way (he says that he forgot the proof). Given this, one can manufacture a number, and try to find its decomposition into a sum of squares. He does that to fall asleep! I think it would keep me awake all night.

Tuesday, February 18th

As always on Tuesdays, no lunch but rather "Journal Club" and later, at 2 p.m., group meeting. I give a little talk on quark number susceptibility during the latter.

At dinner, (roast beef, since it's the next to last dinner for us, and Hans likes it so much) the first subject is the business concerning the visa for Ismail Zahed's new wife, who was denied entry to the US. Gerry called D. Allan Bromley, the President's Advisor for Science and Technology (they know each other from Yale, where Gerry got his Ph.D.) and was assured by him that he cleared the matter up, and that Ismail's wife will be allowed in. Gerry boasts that if that initiative hits a bureaucratic snag, Bromley can always call George Bush, as they also are "chums" from Yale. I guess they call that the "old boys network." Gerry is pleased that this matter is finally cleared up. Next on the list is the primary that was held today in New Hampshire. Hans is worried about Buchanan having obtained 40% of the vote, as this might drive Bush further to the right in order to fend off Buchanan. Tsongas, our collective favorite, won on the Democratic side, though we still believe that he has no chance against Bush.

After Ismail called Gerry to get the latest information, I tell how Ismail was prepared to work on explosion theory for the Algerian government in exchange for his military service. (Since then he received a Green Card and the Algerian government has dispensed his age group from military service.) Hans mentions that there isn't much to do anymore in that field, and that most things are known. Gerry remarks that Hans is the world expert on shock waves and explosions. This gets us to talk about the first test of the bomb. Gerry asks Hans if it was Fermi or Feynman who against regulations stepped outside the trench after the explosion, to measure the strength of the shock wave by dropping little pieces of paper and seeing how far they would be carried by the shock. Hans acknowledges that this was indeed Fermi. Feynman is supposed to also have dodged regulations by looking directly into the flash (though with dark glasses, which everybody was wearing). Hans cannot confirm that, as he said he didn't look at Feynman then. Hans turned out to be at the observing trench roughly 20 km away from the explosion, which took place in New Mexico at "Jornada del Muerte" (literally: "The Journey of Death"), about 100 km away from Los Alamos. Fermi was at the closest trench, at 6 km. Next to Hans was a journalist from *The New York Times*, who was specially selected (as the test of course was Top Secret) and who was supposed to publish an article about the test after the Hiroshima dropping. Roughly a minute after the flash, the shock wave arrived at the trench with Hans and the reporter. When he felt the shock (which was quite substantial, as the yield turned out to be about 20 kilotons) the journalist asked: "What was *that*?" Hans patiently explained it to him. Gerry asks if the 6 km trench was safe, and Hans answers that is was pretty safe, although they didn't know precisely what the yield would be. In fact, there was a betting pool about the yield, and Hans predicted 8 kilotons.

"You were wrong by a factor of 2.5!" Gerry shouts, delighted.

"Well," Hans answers, "I bet the theoretical prediction" (who would have doubted that).

"I obtained the numbers for the neutron multiplication from Bob Serber of the neutron group, and he is a very reliable physicist. He had an assistant, a mediocre physicist, who bet 24 kilotons. Who should I have believed?"

"The pool was finally won by I. I. Rabi, who knew next to nothing about the subject. Fortunately the reporter from the NY Times bet 50 kilotons."

I ask about who was at the test. From the government there were General Groves and two deputies, as well as the head of the nuclear weapons development department James Conant, and Richard Tolman, scientific advisor to Groves. Then there were of course all the division leaders, and most other important people from the project. Secrecy was a big concern, which is why they had it so close to Los Alamos. Nevertheless, the chief of security sighed: "How am I supposed to keep this secret, next time they'll ask me to keep the Mississippi river secret!" Indeed, the test was held in the middle of the night. It was supposed to take place at 3 a.m., however this was cancelled because of rain, as this would have deposited the radioactive material on the ground. The detonation, then, was just before sunrise. The darkness was supposed to help to follow visually the development of the shock wave, which followed immediately after the flash. The flash itself was described by Oppenheimer as "brighter than a thousand suns." By a mechanism that I wasn't able to figure out, the shock front is luminous, ("just like the shock front of a supernova, which we can observe," Hans explained) and was filmed by a high speed camera. The mushroom cloud that emerged glowed in a deep violet first, due to the Cherenkov radiation emitted from the electrons, which were travelling faster than the speed of light in the medium of sand and fission material. Later, the cloud turned grey, as the electrons had slowed down. Then the shock wave hits. As to the secrecy, the Los Alamos community had placed themselves strategically to observe the flash from a hundred kilometers away, and had no problem detecting it. The next morning, there was a story in the "Albuquerque Journal," from a man who was driving home that night, and claimed that he saw the sun rise twice. I ask Hans about the size of the bomb and he spreads his arms:

"About like this, roughly 5 feet wide. They called it 'Fat Man'." Hans mentions that you can now visit the test site, and see a moderately-sized crater, the bottom of which is covered with green glass, the residue of the molten sand. I want more details and ask Hans about the temperature in the middle of the bomb during the explosion.

"This is classified," he announces calmly, "but I can tell you the temperature of the air right outside the bomb, which is available. This is about one million degrees." Cautiously I coax him:

"It is classified, but you know it?"

"Of course," he answers.

"So I suppose that the temperature in the middle of the bomb is higher?" I ask, feeling mildly like a spy. Gerry throws in:

"That is probably classified too!" but Hans signifies no, and says that it is indeed higher. I enquire about how the bomb was attached to the ground. Hans explains that it was attached to a hundred-foot-tall tower, to make the analysis of the shock evolution easier. He describes how Robert Wilson (later with Hans at Cornell) was one of the last ones to check the wiring and the electric connections of the bomb on top of the tower. I ask about Wilson's heart rate at the time. Hans says that it wasn't measured, but that it was very probably considerably elevated. He also says that the tower was pulverized. I ask Hans if everybody was convinced that it worked.

"Most of us were convinced. But there was this little uncertainty about the initiator. After all, Critchfield and I had designed the thing, and ours was in there instead of Fermi's." (I told this story earlier.) The initiator is mainly a source of neutrons that is activated at exactly the same moment as the two shells of fissionable material are blasted together by a conventional detonator. They used an 8 tons detonator (quite a lot, but they wanted to be sure). This was why some of the entries in the pool about the yield were just this number: 8 tons. Those were the pessimists who didn't believe that it would actually ignite. The initiator is mainly polonium, and a little beryllium. I'm not sure if Hans knew that this was declassified, because when I asked him about the initiator, he first hesitated a bit. But I knew about the polonium and told him, he then acknowledged it, and mentioned the beryllium. Whether the beryllium is declassified, I don't know. I ask if there were other tests, and he says that there were none, that next was the Hiroshima bomb, which in fact was a very different design, where a gun would shoot half of the fission material with high velocity on the other half, all this taking place in a tube, which was dropped from a plane. The "shot" was timed such that ignition would take place at a specified altitude.

I ask Hans about a good book which describes the test, and he mentions "The Making of the Bomb" by Richard Rhodes, which Hans says is highly accurate. I mention "Heller als Tausend Sonnen," and Hans remarks that this is a miserable book. In the following, we make fun of the author Robert Jungk.

"He was already a little crazy when he wrote the book, but he got steadily crazier after that!" Hans laughs.

"His book is completely unreliable, you cannot even obtain the truth by believing the opposite of what he says!"

During dessert, we talk about Shakespeare's Hamlet. If anyone is interested in how we made *that* transition, here it is: After all, we talked about polonium, and Laertes' father is, well, Polonius. This is not a joke. Hans compliments me about how well I know Hamlet, and we discuss the Tieck translation. We also argue about whether Polonius was an OK fellow. Gerry maintains that he was just a meddling old fool. I offer that he was treacherous. Anybody could tell we were over dessert, a chocolate pie.

Wednesday, February 19th

No lunch today (Gerry and Hans had an unspecified appointment), so it was on to our last dinner. In the meantime I had bought Rhodes book about the bomb, and brought it to dinner as it had many nice pictures in it, also of Hans and his wife. Looking at those, especially the one of his wife Rose, he exclaims: "How young we were!" There is a photo of General Groves and Oppenheimer at the remains of the firing tower, and Hans mentions that the photo must have been taken weeks after the test, because of the radiation. He also tells how he and Weisskopf drove to the edge of the crater a few days after the test in a jeep. Weisskopf had a Geiger counter, and suddenly went pale when he saw the reading, which indicated 240 rem, as that dose would have led to radiation sickness, and death, in a matter of days, Weisskopf was scared. It turned out, however, that the counter was incorrectly hooked up, and that the correct reading was 250 *minus* what the instrument showed. Considering that these two men are still alive today, with healthy children, it is amazing how people can be scared by exposure to a few *milli*rem. In Rhodes' book, there also was a picture of George Kwiatkowski, riding a horse. He was a chemist, originally from the Ukraine. Hans mentions how, when Kwiatkowski didn't understand something, he would turn to him and ask: "Could you please explain this to a poor chemist?" Hans and Gerry argue about "Who built the bomb" when it comes down to it, "who did most of the work." Gerry says: "I always thought that it was Bacher who really did it." Hans agrees: "Bacher and Kwiatkowski, they did the most."

Earlier, I had read the reference about Hans' initiator, which described it in some detail, and indeed mentioned the polonium and the beryllium, and how it was mixed. It also mentioned a classified aspect of it, namely the shape of the polonium-beryllium mixture, which supposedly was rough like a golf ball, in order to create turbulence in the explosion. When I asked Hans about this feature, he just didn't answer at all.

During dinner, we talked about other things, like how much bigger you chances were to excel in academics if your parents are academics. Hans started to enumerate all the Ph.D.'s in his family, father, grandfathers, cousins etc., but lost count. Also Gerry's father was a University professor, though in first generation. I told a story about Pauli and Heisenberg that I learned from Max Dresden, which amused Hans. Towards the end of their career, Pauli and Heisenberg announce that they have a new unified field theory that explains just about everything. Rumors are flying that they were able to calculate the fine structure constant from first principles. They are to give a joint seminar at Pupin Hall of Columbia University, which is not officially announced, but everybody who is anybody in physics learns about it and tries to attend. Young Max Dresden is there, with standing room only. At the conclusion of the talk, Julian Schwinger raises his hand and comments that their (spin-3/2) theory is blatantly unrenormalizable. After a few more critical comments, the audience is silent, and Niels Bohr, sitting in the front row, slowly gets up. He turns his back to the speakers, faces the audience and exclaims: "It is amazing that these two have ever gotten anything right!"

I ask Hans if it was true that Pauli had palsy, and for that reason was always nodding his head during seminars, as if he agreed with the speaker. Hans vehemently shakes his head. "No, that is completely wrong!" he says, and pulls back his chair a bit.

"He was rocking his whole upper body back and forth! And he always did it, even when he was young! In Copenhagen, where many of the stories about Pauli originated, they made up something about him." And, smiling broadly, he recites:

> Wenn der Pauli mächtig denkt,
> er seinen Oberkörper schwenkt.
> Wenn er mit einer Frage kämpft,
> dann ist die Schwingung ungedämpft![ee]

He tells another Pauli joke that circulated at the time. For introduction, it must be said that Pauli was not very well-liked in seminars, as he always contradicted the speakers. So it is said that when Pauli died, he was received by the angels, who told him that, because he had spent his life searching for the answers to the most difficult problems, and had contributed a lot to the advancement of science, he was allowed to go and see God, and ask him one single question. So he does, and after meeting God, Pauli asks him how to

---

[ee]When Pauli's deep in thought
His upper body rocks.
When fighting with a question,
The oscillation's undamped.

derive the fine structure constant. "Ah," says God, "that's easy!" He goes to the blackboard and starts filling it with equations. After two minutes he filled the board, and presents Pauli with the answer. Pauli looks at the proof, looks at God, and says: "Wrong!"

Maybe because it is the last dinner, the conversation isn't what it usually is. At some point Hans gives me his keys to the apartment, as I will move in after his departure. He retires early. As I will have breakfast with Hans and Gerry the next morning, this was quite all right.

Thursday, February 20*th*

We have breakfast at 8:30. Gerry and Hans talk about physics, as they always seem to do during breakfast. The current problem is the r-process induced by convection and shock waves in the post-supernova environment.

Gerry asks Hans about Oskar Klein, and whether he knew him, and Hans answers yes, that he was the one who introduced him at the Nobel Symposium, and that he was a very nice man. But he adds that he did not know (Theodor) Kaluza, as Klein's name is almost always mentioned along with Kaluza's. After breakfast Hans reads *The New York Times*, as he has been doing for innumerable years. We briefly go down to the patio of the apartment building for some pictures, Hans and Gerry, and then Hans and I. "If it has to be done," says Hans, he doesn't seem to like to be photographed. "For posterity," says Gerry, and Hans answers that he already has his picture in Rhodes' book. Maybe he is just a bit vain, and doesn't like to be photographed at this age. The van that will take Hans and Gerry to the airport has arrived. We briefly shake hands, Hans says that he hopes that we will see each other again, and so do I, hopefully next year. Then I go back to the apartment, alone.

Epilogue[ff]

I have always considered myself extremely lucky that I got to know Hans, this extraordinary man, on such a personal level. We continued these winter meetings at the Kellogg Radiation Laboratory for another eight years, and our relationship grew ever closer. We talked about everything in the world, I was able to teach him some evolutionary biology, and he listened to my work in quantum physics.

[ff]March 4th, 2006.

I never was able to duplicate this feat of writing down all our conversations, but I do have notes from some later years. I also recorded our last few days together in the Spring of 2000, because we all realized that it was possible that it would be our last meeting. Traveling was becoming more and more of a burden for Hans. Unfortunately, these notes were in a bag that was stolen from my office along with my laptop. Everything else that was stolen I was ultimately able to replace, but the loss of these last notes still pains me. But I do recall the last time I saw Hans, back in the same office where I first met him. He was about to leave, and I greeted him at the door. He picked up the cane that he had been using for the last few years, and remarked: "I am more able with my cane," a pun that I had made a few years earlier, and that amused him enough that he repeated it once a year. He was never fond of sentimental farewells, so for our goodbyes he only said two words to me, words that I have been trying to live by ever since. He shook my hand, stronger than the first time we shook, looked me into the eyes, and said: "Carry on."

Acknowledgments

I don't think I have ever properly thanked Gerry Brown for giving me the opportunity to become friends with Hans, and for inviting me to contribute these notes to this volume. I also want to thank Gerry's wife Betty for her friendship in the last twenty years, and particularly Rose Bethe, for the kindness she demonstrated to both me and Taylor, year after year. She took the time to tell me personally that Hans had read these notes shortly before his passing, enjoyed them very much, and encouraged their publication. Finally, I would like to dedicate this diary to my father Nikolaus Adami, who passed away the same year as Hans. He was the one who urged me to keep a diary after I told him that I was going to meet a legend and spend a month with him. "Great men keep diaries!" he encouraged me. He read them with great pride.

Hans Bethe at *The New Yorker*
Jeremy Bernstein

In the fall of 1977, I got the idea of doing a *New Yorker* profile of Bethe. There were several reasons for this, not the least of which was *The New Yorker*'s apparent view of the general energy question. The magazine's mantra on energy was then being presented by Barry Commoner. Commoner, a biologist and noted environmentalist, was a brilliant polemic writer. He made it appear as if "soft energy" was the only path of the righteous. Nuclear energy, for example, was an invention of the devil himself. I was quite sure that many of Commoner's arguments were wrong if examined in detail, and I was equally sure from occasional conversations with the magazine's editor, William Shawn, that there was no interest in any refutation by me. In addition, I was not sure that I knew enough to take Commoner on. I needed an ally and what better ally could there be than Bethe. But I knew there was a problem. In the physics community we all knew Bethe. On every level there was no one we respected more. But Bethe, unlike Commoner, was not a public figure. An article by Bethe, even if Shawn would publish it, which I doubted, would, I thought, not have a great deal of impact. So I invented a scheme. I would do a profile of Bethe two-thirds of which would, if it worked, introduce him sympathetically to the unsuspecting reader who would then be softened up for the final third in which Bethe would explain his views on the general energy question. My idea was that if, in this way, we put Bethe and Commoner on a balance scale, Bethe would tip it. When I proposed doing Bethe's profile to Shawn, needless to say, I did not explain this Machiavellian strategy. Once having gotten Shawn's approval all I had to do was to convince Bethe.

I hardly knew Bethe, but I thought he might be familiar with some of the other profiles I had done for *The New Yorker* such as those of T. D. Lee and

C. N. Yang and I. I. Rabi. From them he might have learned that I was not flake. So I wrote him to ask. He accepted with a caveat ... "I should like to warn you that my personal life is not terribly interesting ... Possibly after the first interview you will come to the conclusion that I am really not very interesting, and in that case please feel free to stop right there." I already knew enough about Bethe's life, including his time at Los Alamos, to know that it was very interesting. So starting in November of 1977, we began a series of tape-recorded interviews, phone conversations and letters that lasted for something like a year and a half. It very soon became clear to me that there were four problems. First was logistics. I was in New York and Bethe in Ithaca. But Bethe came to New York on various sorts of business and I got to Ithaca so that worked itself out. Then there was what I came to call the "feet to the fire" problem. I had explained to Bethe what the plot was, but he had to be reminded periodically. He wanted to cut to the chase and begin debating Commoner at least virtually. I had to keep reminding him of the problem we had and how I was going to solve it by making the reader familiar with him. Then there was the technical competence problem — mine. I was not familiar with some of the science and Hans, as I soon came to call him, was unfailingly patient in explaining and sometimes re-explaining it. Finally, there was the most subtle problem-scoring debating points. By this time Hans had actually debated Commoner and had developed certain catchy phrases. "Wind power," he said, for example "is for the birds." I told him that while that was cute it would lose him sympathy with our readers. It would be much better if he simply explained the limitations, environmental and otherwise, of landscapes festooned with windmills and leave the wise cracks to Commoner. He agreed somewhat reluctantly.

After a year or so of this I began writing what was in length a short book. I turned in the various pieces to Shawn and he accepted them. There then began the process of *New Yorker* editing. This was infinitely meticulous and fortunately a very competent editor Pat Crow took the job. Crow had no scientific training so he had some difficulty with some of the ideas. I recall especially the problem with "watts." He simply would not believe that it was a rate and not an amount of energy. He kept insisting that a seventy-watt bulb burned seventy watts of energy. I forget how I finally convinced him that there was a per second involved. We of course dealt with nuclear energy. But in the middle of this there was Three-Mile-Island. I believe that this really shook Hans. I could tell it in his voice when he called me about it. I said that I thought the only way we could save the piece would be to have a detailed examination of what went wrong. No one was more competent than Hans to do this, and he did. Then we had *The New Yorker* checkers — two people who were assigned to check every fact in the piece. It was now scheduled to run in three parts beginning on December 3, 1979. There was so much work to be done that I had given up my Christmas vacation plans to be around to answer questions. Then came the thunder bolt — Shawn had pulled the piece.

At first I was not too disturbed. *New Yorker* scheduling often involved last-minute changes. Sometimes a new article had come in that was so timely that it needed to be put into the magazine right away. There were also space issues. A three-part article required a lot of space and if the magazine was shrunk because of less advertising the article might be postponed. I asked the checker who had called me with the news to see if my article had been re-scheduled for a later set of issues. He said he could not find any re-scheduling. That started to worry me. I decided to ask my friend and former editor Gardner Botsford if he knew anything. Botsford was second in command at the magazine. He said that Shawn had come into him with some article on animals that he had taken from the "bank." *The New Yorker* had a huge bank of unpublished articles which often sat around for years before they found their way into the magazine. Shawn wanted Botsford to edit it to be put into the issue that was supposed to start the Bethe series. Now I knew that something was seriously wrong and I began to boil. Bethe and I had been working on this for nearly two years and for Shawn to do something like this without any explanation was unforgivable. By now I had been writing for the magazine for nearly twenty years and during that time I had never voiced a complaint. I felt myself lucky to be able to write for such a great magazine. But this? This was beyond the pale. In a towering rage I stormed into Shawn's secretary May Painter and explained what had been done and said I would never write another word for the magazine. She had never seen me like this and, in some state of alarm, after I left she went in to see Shawn. The next day the piece was back in the schedule. Shawn never apologized and never explained but it was clear to me that Three-Mile-Island was the last straw for him. This is how the profile of Bethe got into *The New Yorker*.

About a year before he died I spoke to Hans for the last time. This was also about some writing I was doing. I was writing a brief biography of Oppenheimer in the spirit of a *New Yorker* profile. I had come across something that I thought Hans could help me with. In January of 1954 there had been a Rochester Conference in Rochester which I attended. I had kept the published minutes. I noticed that there was a session that Oppenheimer presided over during which he asked Teller for some clarification of some experimental results. I knew that a couple of months later the Oppenheimer hearings began and I wondered if Hans and Oppenheimer had discussed these matters that January. Hans said that indeed they had and that Oppenheimer had told him that he was going to lose. I asked Hans how he was. He said that physically he was not in very good shape. "But," he added, "I can still think."

My Sixty Years with Hans Bethe

Edwin E. Salpeter

I first met Hans Bethe in 1946 in Birmingham, England, when I was a starting graduate student, but I had already read a couple of his papers two years earlier as an undergraduate. I am making a point of Hans having had an influence on me for well over sixty years because of my somewhat unorthodox (and brash) aim for this essay: Instead of writing about Hans's publications, I will mainly write about my own publications and how they were affected by Hans, sometimes in subtle ways. I will also mention the importance of "academic grandfathers," i.e. the teachers and mentors of one's own teachers, advisors and role models. This group can include a small number of people in addition to the thesis advisor, which can lead to some seeming contradictions and dichotomies.

Although I was not aware of it at the time, I already met such dichotomies in two forms when I was doing a Ph.D. thesis at Birmingham University with Hans's close friend Rudi Peierls. My own thesis was to be in quantum electrodynamics, but many of Rudi's other graduate students worked in very different fields. When Hans visited Rudi's department he would go from student to student, listening first and then giving advice, to return again half a year or a year later — just like a chess master playing multiple boards. The first impression of Hans then was his amazing ability to understand all fields of physics and to remember previous discussions and problems. More important, however, was a seeming schizophrenia: He advised rigorous mathematical techniques for some of the student's problems, but the opposite, i.e. qualitative approximations, for others. Added to this was his delight in not only giving technical advice but also advice on social questions about each student's future. I was very pleased with Rudi Peierls as a thesis advisor and Hans Bethe was similar but even more so, with his help to students who were not even his own. I knew I wanted to come to Cornell to be Hans's postdoc, which I did in 1949 — and I have been there ever since.

Hans Bethe's multi-faceted abilities stem in part from the disparate teachers he had had himself, especially his own thesis advisor Arnold Sommerfeld in Munich and Enrico Fermi in Rome. This background is described more beautifully in Sam Schweber's article here and in his book,[21] but can be summarized briefly. Sommerfeld taught him how to use rigorous, even if complicated, mathematical techniques for a quantitative expose of a whole branch of theoretical physics. Fermi, on the other hand, taught him how to make major short-cuts towards approximate results and to utilize the light touch of intuition especially for a topic with uncertainties. These seem like diametrically opposed teachings and it was Hans's own genius which enabled him to know when to use one and when the other (and occasionally even to combine them). His advice to us youngsters with less intuition was "learn advanced mathematics in case you need it, but use only the minimum necessary for any particular problem." In addition to his two major teachers, Hans was also influenced strongly by Paul Ewald in Stuttgart, where he stayed in 1929. Ewald taught him that physics should and can be fun, as can comradery among physicists. In a similar vein, ten years later Hans married Ewald's daughter Rose, who has been an inspiration to Hans and to us youngsters ever since. Hans's continuing friendship and collaboration with Rudi Peierls may be partly due to both of them, as well as Paul Ewald, having been Sommerfeld's students. Similarly, three of the authors in this book, Dyson, Brown and myself, were students of both Rudi and Hans.

The confusing three eras of quantum electrodynamics, as a continuation of modern quantum mechanics, are described in Freeman Dyson's article. The first era, before 1947, incorporated Dirac's modern relativistic quantum mechanics in the interaction of electrons and photons and was rather complicated and beset by infinities. The second era was a short-lived revolution in 1947, brought on by the now famous Shelter Island Conference, where the experimental discovery of the Lamb shift was announced. This shift in two energy levels relative to each other contradicted results from the Dirac equation and presented a challenge to quantum electrodynamics. Hans provided an explanation after first inventing an approximate and simple early version of renormalization theory. His paper[5] was so simple (and non-relativistic) that it was essentially written on a train ride back from the conference. This paper did away with the singularities of the first era and his simple calculation gave a good approximation to the experimental value of the Lamb shift — pure Fermi rather than Sommerfeld! The third era returned to the fully relativistic theory, to self-consistency and rigor, and was initiated by Feynman, Schwinger and Tomonaga and consolidated by Dyson.

I can attest to Hans's influence on each of these three eras from my own history: Hans had written a *Handbuch der Physik* compendium in 1933 on the quantum mechanics of one- and two-electron atoms, which was the forerunner of the first era in 1933. Twenty-five years later I was involved in bringing this book up to date and publishing it in English.[11] It was fun writing that book (and we still sell one or two copies per year), but it was remarkable how few changes were necessary from the original version in spite of the 25 years — Hans had both displayed Sommerfeld's thoroughness and also anticipated so much! I can illustrate the importance of Hans's invention of the second era of quantum electrodynamics from my own Ph.D. thesis which involved a rigorous and tedious analysis of the singularities encountered in the formulation of QED during the first era: My thesis and paper were correct and I still got my Ph.D. in 1948, but they were completely out of date by the time my paper came out, so I did not even bother to quote it in my autobiography. My example from the third era is a bit more confusing:

In 1951 Hans and I published a paper on a fully relativistic quantum electrodynamics equation for treating bound states between two Fermi-Dirac particles.[20] This was motivated in part by a suggestion for the massless photon being the bound state of two particles with mass but with such a strong attractive potential that the potential energy cancels the rest mass energies exactly. This would require a fully relativistic treatment and, in keeping with relativistic four-space, we had to introduce a separate time-variable for each of the two particles to go with the two separate spatial coordinates. Our equation fitted in nicely with the Feynman formalism and was quite elegant, but the two separate times presented great difficulties in understanding and applications.

This difficulty is illustrated by an apocryphal story of a theoretical physics graduate student confiding in Hans that he had tried to study the Bethe–Salpeter equation but had not really understood it, with Hans answering "don't worry — neither had I really." That lack of understanding also went for me in spades and we never managed to make much practical use of the equation ourselves for a very strongly bound pair of particles (although there were some advances, described in References 13 and 15). However, I was able to apply Hans's motto of compromise and use of successive approximations in atomic physics: The hydrogen atom is mostly a non-relativistic bound state, but the small energy level splittings in fine structure and hyperfine structure had been measured to very great accuracy. For the QED theory of these splittings one had to calculate relativistic corrections to the motion of the more massive nucleus. I was indeed able to use the Bethe–Salpeter equation to calculate these corrections to quite good accuracy.[17]

Hans might well have gotten a Nobel Prize for his 1947 Lamb shift paper, which revolutionized quantum electrodynamics. In fact, he got it instead for three articles on the conversion of hydrogen into helium in the sun which he had written almost ten years earlier. These had revolutionized the theory of stellar energy production and started the science of nuclear astrophysics.[3,4,9] This work is described briefly in the article with John Bahcall in this volume, but I got involved in nuclear astrophysics only in the early 1950's, largely because of Hans's work on nuclear physics theory in the late 1940's and thereafter. In particular, Hans had just invented a practical version of "effective range theory" for nuclear scattering. In true Bethe fashion this theory eliminated the necessity of specifying an explicit nuclear potential function and related scattering experiments to just two empirical constants.[6,10] Although Hans and I were still involved in quantum electrodynamics for some time later, he also put me onto some nuclear theory problems in 1950, including some improvements to nuclear effective range theory. As he mentions in his autobiography,[7] Hans was very helpful to Willy Fowler at Caltech, who had a long-range program of nuclear experiments to elucidate nuclear reactions and energy production in stars. Since combining Hans's theoretical know-how and Willy's experimental results would be very useful, Hans sent me, at least as a minor surrogate, to do theoretical calculations in Willy Fowler's lab for the summer of 1951.

Already in 1939, Hans had looked forward to improving the accuracy of the proton-proton chain calculations and to investigate further reactions in evolved stars which had exhausted their hydrogen. I started to look at these two topics in the summer of 1951 and by the end of the summer I was ready to write a long paper on each of these topics. I like bragging about having produced two such calculations in a single summer (in spite of distractions from the Bethe–Salpeter equation), but this success was not really due to my being intrinsically fast but due to all the ground work that had been done by Hans. To start with the proton-proton reaction to form a deuteron, an electron and a neutrino: Because of Hans's effective range theory I did not have to evaluate any matrix elements from an assumed nuclear potential function (apart from not knowing what the right function was anyway). Accurate proton-proton scattering experiments had been carried out by then, which gave the effective range theory constants and the reaction rate with almost no calculations at all. This was not only simple but also gave a reliable estimate of the accuracy of the calculation which by now was pretty good. Although Hans had omitted the He^3 plus He^3 reaction as one possible completion of the chain in 1938, by 1951 this reaction had been suggested by others and experiments had been done for the sister reaction of tritium-tritium scattering. I therefore was also able to calculate this reaction rate simply and reliably.[18]

Hans already knew in 1938 that building up heavier elements from helium and producing energy in evolved stars would be complicated since two-body reactions of helium with hydrogen or another helium nucleus do not produce a stable nucleus. I consequently had to look at many possibilities in the summer of 1951, but Hans had made it easy for me to calculate out each possibility. An enormous amount of prescriptions were in the part of the "Bethe Bible" on nuclear dynamics[2] and his C-N-cycle paper already included prescriptions for calculating three-body reactions. It became clear pretty quickly that one such three-body reaction, the "triple alpha process" to convert three helium-4 nuclei into one C-12 nucleus, would do the trick. Fortunately, Willy Fowler was able to tell me that Be-8, although not a stable nucleus, has a resonance level at an appropriate energy so that a little of Be-8 is always in equilibrium with alpha particle pairs. This made the calculation easier in terms of two two-body reactions and, more important, made the overall reaction rate very much larger than without the resonance, but still very sensitive to both temperature and density. By 1951 there were already mathematical models for the interiors of red giant stars, although they were still quite approximate. These models reached quite large central values for both temperature and density in the later stages of red giant evolution. Even though my calculation had no rate enhancement from any resonance levels in C-12, since none were known experimentally in 1951, I was able to show that stars at the tip of the red giant branch could start a new era of energy production from helium burning near the center.[19] This possibility for evolved stars was in contrast to the early cosmological stages of the Big Bang expansion where high temperatures could also be reached but densities were much too low for helium burning. This settled one long-standing controversy about Big Bang versus stars for making heavy elements and it is not surprising that I got tenure as an associate professor within a couple of years.

Unfortunately for my later reputation this was not the end of the helium burning story, because I had failed to follow another one of Hans Bethe's tenets: "You should look at ALL the experimental information at hand, not only the most relevant, and be prepared to make conjectures if that helps." With my assumption of no resonance level in C-12 my required temperature was high enough so that the rate of the next reaction, C-12 plus He-4 making O-16, would be very rapid. As a consequence, the predicted abundance ratio of oxygen to carbon would be very large. However, it was already known in 1951 that the two abundances were comparable in the universe, a fact also useful for biology. A resonance level at an appropriate energy in C-12 would greatly increase the reaction rate at a given temperature and hence lower the required central temperatures for helium burning slightly, which lowers the rate of converting carbon to oxygen. Furthermore, it is not that such a resonance level was known to be absent in 1951, but merely that it had not yet been looked for experimentally and yet I did not follow Hans's advice of making conjectures. Fred Hoyle, on the other hand, not only made such a conjecture just a few years later, but predicted at what energy the new resonance level in C-12 should occur to have the calculated O/C abundance ratio agree with the observed one. Willy Fowler's experimental group soon found the resonance level just where it should be, which started a long Fowler/Hoyle collaboration.

As my helium-burning fiasco illustrates, questions on stellar energy production or loss are related to, but different from, those on abundances of all the isotopes of all the elements. The abundance questions were resolved satisfactorily over the years by a team built around Burbidge, Fowler and Hoyle plus a smaller team around Al Cameron. I concentrated more on the ins of energy from nuclear build-up and the outs of energy from neutrino loss and inverse beta-decay. These were questions which received encouragement and advice from Hans, but also involved building models for the interiors of very highly evolved stars. Such stars have more disparate layers with different composition and even higher central densities than mere red giant starts and needed electronic computers for some of the model building. Some such model building was done in the 1960's and 1970's by my graduate students, post-docs and younger colleagues, including John Cox, Willy Deinzer, Gilles Beaudet, Bob Malone, Jonathan Katz, Alan Boozer and Paul Joss. The same time-period saw calculations of energy loss from neutrino emission, mentioned in the Bahcall article, and early considerations of solar model complications related to solar neutrinos by Giora Shaviv. All these calculations were beset by complexities and my major (or only) contribution was to insert Hans Bethe's admonition to make, for the sake of simplicity, the greatest approximations that one could just barely get away with. Hans was and is an "academic grandfather" for all these youngsters in

References

1. APS Study Panel, "Science and Technology of Directed Energy Weapons," *Rev. Mod. Phys.* **59**, 51 (1987).
2. H. A. Bethe, *Rev. Mod. Phys.* **9**, 69 (1937).
3. H. A. Bethe, *Phys. Rev.* **55**, 434 (1939).
4. H. A. Bethe, *Astrophys. J.* **92**, 118 (1940).
5. H. A. Bethe, *Phys. Rev.* **72**, 339 (1947).
6. H. A. Bethe, *Phys. Rev.* **76**, 38 (1949).
7. *Annu. Rev. Astron. Astrophys.* **41**, 1 (2003).
8. H. A. Bethe and J. Ashkin, in *Experimental Nuclear Physics*, ed. E. Segrè, John Wiley, New York, 1953.
9. H. A. Bethe and C. L. Critchfield, *Phys. Rev.* **54**, 248 (1938).
10. H. A. Bethe and C. L. Longmire, *Phys. Rev.* **77**, 647 (1950).
11. H. A. Bethe and E. E. Salpeter, *Quantum Mechanics of One and Two Electron Atoms*, Springer Verlag, Berlin, 1957.
12. L. Bildsten, E. E. Salpeter and I. Wasserman, *Ap. J.* **384**, 143 (1992).
13. R. E. Cutkosky, in *Perspectives in Modern Physics*, ed. R. E. Marshak, Interscience, New York, 1956.
14. B. T. Draine and E. E. Salpeter, *Ap. J.* **231**, 77 (1979).
15. J. S. Goldstein, *Phys. Rev.* **91**, 1516 (1953).
16. R. W. Nelson, *Science and Global Security* **10**, 1 (2002).
17. E. E. Salpeter, *Phys. Rev.* **87**, 328 (1952).
18. E. E. Salpeter, *Phys. Rev.* **88**, 547 (1952).
19. E. E. Salpeter, *Ap. J.* **115**, 326 (1952).
20. E. E. Salpeter and H. A. Bethe, *Phys. Rev.* **84**, 1232 (1951).
21. S. S. Schweber, *In the Shadow of the Bomb: Bethe, Oppenheimer, and the Moral Responsibility of the Scientist*, Princeton Univ. Press, Princeton, 2000.
22. G. Shaviv, in *The Initial Mass Function 50 Years Later*, eds. E. Corbelli, F. Palla and H. Zinnecker, Springer, Dordrecht, 2005.
23. D. J. Stevenson and E. E. Salpeter, *Ap. J. Suppl.* **35**, 239 (1977).
24. E. E. Salpeter and H. M. Van Horn, *Ap. J.* **155**, 183 (1969).
25. H. S. Zapolsky and E. E. Salpeter, *Phys. Rev.* **158**, 876 (1967).

Hans Bethe*
Kurt Gottfried

There are a handful of people who soar, whose accomplishments are so off-scale as to nearly defy belief. Hans Bethe (2 July 1906–6 March 2005) was of that caliber. As just one measure of his stature, imagine the task of copying his published opus by hand, for that is how he wrote most of it; an industrious scribe would labor for many years without even tackling his innumerable government studies and reports. This quantitative dimension would be unremarkable were it not that the mountain of paper would contain seams of previously unknown intellectual gold, nuggets of ingenious invention, brilliant syntheses of new knowledge, technical analyses that altered the geopolitical landscape, and calls to keep morality in mind.

This issue of *Physics Today* breathes life into the preceding paragraph with articles that sketch Hans's contributions to most, but certainly not all, of the areas in which he played such seminal roles for more than 70 years. Appearing in roughly chronological order.

- ▶ Silvan Schweber writes on the period before World War II — Hans's education, swift rise to international prominence, emigration to the United States, and impact on American physics;
- ▶ The late John Bahcall and Edwin Salpeter discuss Hans's work on energy production in stars, nuclear astrophysics in general, and neutrino physics;
- ▶ Freeman Dyson relates the history of Hans and quantum electrodynamics, both before the war and, later, after the discovery of the Lamb shift;

---

*Guest editor's Introduction of Special Issue: Hans Bethe, *Physics Today*, October 2005, pp. 36–37. Reprinted with permission, © American Institute of Physics.

- Richard Garwin and I describe Hans's long involvement with national defense — the Manhattan Project, the hydrogen bomb, the cold war, and arms control*;
- John Negele tells of Hans, the nuclear many-body problem, and nuclear matter; and
- Gerald Brown gives us an intimate look at his remarkable collaboration with Hans on supernovae and other astrophysical topics, in the last decades of Bethe's life.

We provide an overview of Hans's accomplishments and offer glimpses of the talents and erudition that made it all possible. These talents and erudition can easily be inferred from the scientific literature itself. But that literature cannot describe his personality and character. The authors of these articles, all having had the good fortune to enjoy long and unforgettable friendships with Hans, have sought to offer some insights into how it came to be that he had so special a place in the hearts and minds of his colleagues and of the community at large — a respect and esteem that are not explicable solely in terms of his tangible contributions to scientific knowledge and the public good.

A Fuller Picture

In an effort to give a fuller picture of the man, I add a few words to what you will find in the articles that follow.

Hans loved nature and the outdoors — he was a very strong man physically, and until late in life, he hiked over distances and at altitudes that few of his contemporaries could match. At Cornell, he would take a roundabout route to lunch if the flowers were in bloom. But his self-discipline was daunting — in a discussion in his office he would always be fully focused on the topic at hand, but if you looked over your shoulder as you left he would already be writing on his stack of paper. If the air conditioning broke down in the heat of summer, he would have a fan brought in.

Hans had a great sense of humor and a resonant laugh. He made sure that his *Selected Works* (World Scientific, 1997) included the "scandal" of his youth — the spoof of Eddington described by Sam Schweber in his article. And late in life he liked to quote Justice Oliver Wendell Holmes's line "Oh, to be 80 again." With a twinkle in his eye, he once told me that he

---
*This article is not reprinted in this volume.

Hans in the Alps, where he would go for a week or two every summer before he emigrated to the US. This photo is circa 1930. (Courtesy of Rose Bethe.)

had just told Senate staff that he could not testify at an important hearing if it were postponed by a few hours, because of a prior commitment; what he had not told them was that the engagement was a dinner with his former student Leonard Maximon. Students at all levels — from high school to Ph.D. candidates — were a high priority for him and they always found him available.

Hans was an unassuming man — anything but a prima donna. Once during lunch with him and Victor Weisskopf, the conversation led Viki to say that a particular physicist was habitually arrogant toward his colleagues, and I agreed. Hans was astonished — "Gosh! He's never been that way with me." Viki and I just looked at each other silently, and then Hans blushed a bit and quietly cleared his throat.

Nevertheless, Hans had an objective evaluation of his off-scale talents and strengths, combined with a clear recognition that he was not a magician like Paul Dirac. And he was competitive — he once told a younger colleague that one should only work on problems for which one had an unfair advantage!

To the end, he was seeker of new knowledge, devoted to actually doing science. Well into his nineties, in conversations about his work with Gerry Brown, he continued to display a stunning command of physics — his strong voice moving steadily within a few sentences from shock waves to neutrino interactions to thermodynamics.

For almost seven decades, his wife Rose was his constant companion and closest adviser. Together, they explored the Rocky Mountains in the summers before Pearl Harbor. In the difficult decisions that Hans faced during the controversies surrounding the H-bomb, it was primarily with Rose that he discussed how to deal with the unprecedented moral dilemmas that arose. In the last years of his life, as his physical strength ebbed, it was Rose and their son Henry who kept him abreast of the political scene, which distressed him deeply, and who made it possible for him to continue to participate in public policy debates and to satisfy his unquenchable curiosity.

Hans in his office at Cornell, circa 1935–1936. (Courtesy of Rose Bethe.)

Hans and Rose Bethe in the Canadian Rockies in the summer of 1959. (Courtesy of Rose Bethe.)

"The Happy Thirties"*
Silvan S. Schweber

Hans Bethe was born on 2 July 1906 in Strasbourg, when Alsace was part of the Wilhelminian empire. His father, Albrecht Bethe, was trained as a physician and became a widely respected physiologist. In 1915 he accepted a professorship in the newly founded Frankfurt University. Hans's mother, Anna, was raised in Strasbourg where her father was a professor of medicine. An only child, Hans grew up in a Christian household, but one in which religion did not play an important role. His father was Protestant; his mother had been Jewish but she became a Lutheran before she met Hans's father. Hans's mother was a talented and accomplished musician, but a year or two before World War I her hearing was impaired as a result of contracting influenza. The illness left psychological scars, and she became prone to what was diagnosed at the time as bouts of "nervous exhaustion," or extended periods of depression. The marriage suffered under the strain, and Hans's parents eventually divorced in 1927. From the mid-1920's on, it was Hans who looked after his mother's well-being.

One of Bethe's earliest memories was being interested in numbers and playing with numbers. His numerical and mathematical abilities manifested themselves early. His father told of Hans at age four sitting on the stoop of their house, a piece of chalk in each hand, taking square roots of numbers. By the age of five, he had fully understood fractions and could add, subtract, multiply, and divide any two of them. At age seven he was finding ever-larger prime numbers and had made a table of the powers of two and of three, up to 2^{14} and 3^{10}, and had memorized them. At age fourteen he taught himself the

---

*Reprinted with permission from *Physics Today*, October 2005, pp. 38–43. © American Institute of Physics.

calculus by reading Walther Nernst and Arthur Schönflies's *Einführung in die mathematische Behandlung der Naturwissenschaften* (Introduction to the Mathematical Treatment of the Natural Sciences), which he had "stolen"[1] from his father's library and read on the sly.

Bethe started reading at the age of four and began writing in capital letters at about the same age. Very soon after mastering the art of handwriting, he began filling large numbers of little booklets with stories. His mode of writing was distinctive: He would write one line from left to right and the next line from right to left! Many years later while visiting Crete, he was pleased to learn that the Greeks wrote the same way on their tablets in 700 BC: in capital letters, left to right and then right to left.

Though somewhat sickly as a young boy and frequently absent from school, Bethe nonetheless was an outstanding student. His mathematics teacher in Frankfurt recognized his outstanding mathematical talents and encouraged him to continue studies in mathematics and the physical sciences.

A young Hans Bethe. (Courtesy of Rose Bethe.)

By the time Bethe finished high school (gymnasium) in the spring of 1924 he knew he wanted to be a physicist because "mathematics seemed to prove things that are obvious." After completing two years of studies at Frankfurt University, he had exhausted the resources in theoretical physics in Frankfurt and was advised to go to Munich. In the summer of 1926, he joined Arnold Sommerfeld's seminar in Munich.

From Sommerfeld to Fermi

Sommerfeld was a forceful and charismatic figure,[2] and although he was very much the *Herr Geheimrat* (literally "Privy Councillor," an exclusive, greatly respected honorary title bestowed on civilians by the government), the atmosphere of the seminar was nonetheless characterized by the intellectual give-and-take between him and his students and assistants. In contrast to the usual practice at other German universities, where only invited guests spoke, Sommerfeld had his students and assistants make presentations in his seminar. Thus, shortly after coming to Munich, Bethe reported on Schrödinger's paper on perturbation theory in wave mechanics.

It was in Munich that Bethe anchored his self-confidence. There he discovered his remarkable talents and his exceptional proficiency in physics. Sommerfeld told him that he was among the very best students who had studied with him. His self-confidence in physics quickly extended to other matters. At a symposium held in October 1988 to mark the 80th birthday of his friend Victor Weisskopf, Bethe was introduced by Kurt Gottfried, who narrated the following story. In 1934 Weisskopf told Bethe that he was about to undertake a calculation of pair production for spin-0 particles, a calculation similar to one that Bethe had performed the previous year for spin-1/2 particles. Weisskopf wanted to know how long it would take to do the computation. Bethe answered, "Me it would take three days; you three weeks." At the start of his talk, Bethe commented, "I was very conceited at that time. I still am — but I can hide it better."

Bethe obtained his doctorate summa cum laude in 1928 with a thesis on electron diffraction in crystals. In his thesis he explained why electrons within certain energy intervals were observed to be totally reflected.[3] Building on previous work by Paul Ewald on the diffraction of x-rays by crystals, and making use of the fact that the electron's wavefunction inside the crystal must be of the form $\exp(i\mathbf{k}\cdot\mathbf{r})\,u_k(\mathbf{r})$, with $u_k(\mathbf{r})$ having the periodicity of the crystal lattice, Bethe established that for certain incident directions and energy intervals there did not exist any wavefunctions corresponding to an electron propagating in the crystal. The calculation was a difficult one, and the connection between the forbidden intervals and the gaps between the energy bands of electrons in metals was not recognized until later, after Felix Bloch's work had received wide acceptance (see Reference 3).

In the fall of 1929, Sommerfeld recommended Bethe for a Rockefeller Foundation fellowship. And so during 1930 Bethe spent a semester in Cambridge under the aegis of Ralph Fowler, and a semester in Rome working with Enrico Fermi.

Bethe found the openness and cordiality of his British hosts, Fowler and Patrick Blackett in particular, most engaging and attractive. Clearly the Cambridge surroundings allowed some of the stiffness that a German education bestowed on scholars to be shed, for the year 1931 opened with an astonishing short original article (*Kurze Originalmitteilung*) entitled "On the Quantum Theory of the Temperature of Absolute Zero" in the journal *Naturwissenschaften*. The paper was signed by Guido Beck, Bethe, and Wolfgang Riezler, three postdoctoral fellows at the Cavendish Laboratory. Coming on the heels of Arthur Eddington's attempt to explain the numerical value of the fine structure constant, the article pretended to give an alternative derivation of its value.[4] Since papers in respected scientific journals in those days were read with absolute trust in the honorable intentions of the authors and editors, it took a while for the community to realize that *Naturwissenschaften* had been had and that the paper was a prank. Arnold Berliner, the editor of *Naturwissenschaften*, was not amused. Nor was Sommerfeld. Berliner demanded an apology and on 6 March a "correction" appeared in the journal.

From Cambridge, Bethe went to Rome (see his reminiscences in *Physics Today*, June 2002, page 28). After a few weeks there he wrote Sommerfeld

> 9 IV 31
>
> The best thing in Rome is unquestionably Fermi. It is absolutely fabulous how he immediately sees the solution to every problem that is put to him, and his ability to present such complicated things as quantum electrodynamics simply .... I am now actually sorry that I cannot stay here longer, or as the case may be, that I did not come here for all of the Rockefeller-time.[5]

Bethe returned to Rome in the spring of 1932, but in the meantime he had obligated himself to write two lengthy reviews for the new edition of the *Handbuch der Physik* that Adolf Smekal was editing. One, on electrons in metals, was to be written with Sommerfeld, and the other was to present the state of knowledge of one- and two-electron atoms.

In Rome, Bethe was exposed to the much freer and more informal mode of interaction between Fermi and his students than what he had experienced in Munich. Though only five years older than Bethe, Fermi became — besides Sommerfeld — the other great formative influence on him. Fermi helped Bethe free himself from the rigorous and exhaustive approach that was Sommerfeld's hallmark. From Fermi, Bethe learned to reason qualitatively, to obtain insights from back-of-envelope calculations, and to think of physics as easy and fun, as challenging problems to be solved.

Bethe's craftsmanship was an amalgam of what he learned from these two great physicists and teachers, combining the best of both: the thoroughness and rigor of Sommerfeld with the clarity and simplicity of Fermi. This craftsmanship is displayed in full force in the many reviews that Bethe wrote. His first, the result of Sommerfeld's asking him to collaborate in the writing of his *Handbuch der Physik* entry on solid-state physics, exhibited his remarkable powers of synthesis. Of his reviews, the two *Handbuch*[6] entries and the "Bethe Bible" — three articles on nuclear physics[7] in *Reviews of Modern Physics* in 1936 and 1937 — were the most famous. Calling Bethe's reviews "reviews," however, is a misnomer. They were syntheses of the fields, giving the subjects coherence and unity and charting the paths to be taken in addressing new problems. They usually contained much that was new, material that Bethe had worked out in preparing the essay.

1933 and Its Aftermath

Already as a teenager, and throughout the 1920's, Bethe kept abreast of the political and economic developments in Germany. As attested to by his letters to Sommerfeld from 1928 on, finding suitable employment and making ends meet were constant worries. In the fall of 1932, Bethe obtained an appointment in Tübingen as an acting assistant professor and started teaching theoretical physics there. But Adolf Hitler's rise to power on 30 January 1933 changed all that.

In April 1933, shortly after the enactment of the racial laws which forbade any Jew, half-Jew, or quarter-Jew from holding any state or federal governmental position, Bethe lost his job. Sommerfeld was able to help by awarding Bethe a fellowship in Munich for the summer of 1933. He also got William Lawrence Bragg to invite Bethe to come to Manchester for a year to replace Evan James Williams, who was going to Copenhagen to work with Niels Bohr.

The warm relationship between Bragg and Sommerfeld dated back to the 1910's, after the discovery of x-ray diffraction at Sommerfeld's institute. It had become even closer in the early 1930's when Sommerfeld helped Bragg recover from a bout of deep depression. Bethe conveyed his impression of Manchester in the first letter he wrote to Sommerfeld after arriving there:

> 23 XI 33
>
> The best thing in Manchester are the people at the Institute. Bragg is marvelous, humanly and physics-wise .... He makes very interesting experiments about the arrangement of atoms in alloys (superlattices) and I attempt to devise theory for them. It is a pleasure to tell him things: He understands all essential points in the shortest time, while that is mostly very difficult with experimenters ....[5]

The appointment in Manchester was for a year, and thus the question of what would happen the following year came up early on. There then occurred a confluence of events that determined Bethe's subsequent life. Cornell University was looking for a theorist, and Lloyd P. Smith, a young theorist there who had studied with Bethe in Munich, recommended him strongly for the position. At the same time, Bragg was visiting Cornell for the spring semester and could corroborate Smith's assessment of Bethe.

On 18 August 1934, R. Clifton Gibbs, chair of the Cornell physics department, wrote Robert M. Ogden, the dean of arts and sciences, recommending the "appointment of Dr. Hans Bethe as Acting Assistant Professor of Physics for the year 1934–1935, at a salary of $3000":

> The strong recommendations of Profs. Sommerfeld and Bragg and our intimate knowledge of the admirable way that he exerted his influence in promoting the work in theoretical physics at the University of Munich together with his numerous outstanding publications (a list of which is attached) have convinced a large majority of the Faculty in Physics that Dr. Bethe is a most promising candidate in meeting our needs.[8]

Bethe accepted, but because he had received an offer of a yearlong fellowship in Bristol with Nevill Mott, he asked and obtained permission from Cornell to assume his duties there in the spring term rather than at the beginning of the academic year. He stayed in Bristol during the fall semester of 1934 and arrived in Ithaca, New York, in February 1935.

Lothar Nordheim, Hans Bethe, I. I. Rabi, and Edward Condon in the mid-1930's. (Courtesy of AIP Emilio Segrè Visual Archives.)

Cornell University

When Bethe joined the physics department at Cornell, it consisted of some 15 faculty members, and about 40 graduate students were enrolled. He very soon felt "quite at home." When he went back to Germany that summer to visit his mother, he had become convinced "that probably [he] would remain at Cornell for a long, long time." In the fall of 1935, Robert Bacher joined the department. Bethe, Bacher, and M. Stanley Livingston made Cornell into an outstanding center of nuclear research. Although Cornell's cyclotron only produced 1.2-MeV deuterons, Livingston and his associates developed an arc source that transformed the cyclotron into a particularly useful tool for neutron research. Bethe not only provided suggestions for experiments and the theory for their interpretation, but was intimately involved with their design and data analysis. At Cornell, like at the other centers where nuclear physics was being cultivated, theorists and experimenters worked closely together. At the beginning of the 1936–1937 academic year, Bethe confessed to Sommerfeld that although he had gone to Cornell with mixed feelings,

> like a missionary going to the darkest parts of Africa in order to spread there the true faith ... already half a year later I no longer held this opinion and today I hardly would return to Europe even if I would be offered the same amount of dollars as at Cornell.
>
> The characteristic trait of physics in America is team work. Working together within the large institutes — in every proper one everything that physics encompasses is being done — the experimentalist constantly discusses his problems with the theorist, the nuclear physicist with the spectroscopist. By virtue of this cooperation many of the problems are immediately disposed of, [whereas] that would take months in a specialized institute. More team work [in English in the original]: the frequent conferences of the American Physical Society ... .[5]

The influence of the émigré scientists who had come to the US was particularly noticeable at the many theoretical conferences organized to assimilate the insights that quantum mechanics was providing in many fields, especially molecular physics and the emerging field of nuclear physics. The Washington Conferences on Theoretical Physics, initiated in 1935 by Merle Tuve and John Fleming of the Carnegie Institution, jointly sponsored by the Carnegie Institution and George Washington University and held annually until 1942, were paradigmatic of such meetings. Their intellectual agenda was set by George Gamow and Bethe's friend Edward Teller. Their purpose was to evolve in the US something similar to the Copenhagen Conferences, in which a small number of theoretical physicists working on related problems would assemble to discuss in an informal way the difficulties they had met in their research. The conferences proved to be extremely influential and seminal, partly because they were restricted to theory and partly because their size was strictly regulated so they could remain working conferences (see the table on page 140).

Bethe attended the 1935 and 1937 Washington Conferences; when invited to the 1938 conference, though, he indicated to Teller that he was not interested in the problem of stellar-energy generation. It was only after Teller's repeated urgings that Bethe agreed to attend. The subject of the conference had been suggested by Gamow, who in 1938 was ideally positioned to solve the problem of energy production in stars. He recognized the interrelation of nucleosynthesis and energy production, and together with Edward Teller

Participants at the 1938 Washington Conference on Theoretical Physics, the fourth of a series sponsored by the Carnegie Institution of Washington and George Washington University. The conference introduced Bethe to the problem of stellar-energy generation; his definitive solution to that problem earned him the 1967 Nobel Prize. (Courtesy of Special Collections and University Archives, George Washington University, Washington, D.C.)

The Washington Conferences 1935–1942*

Date	Topic	Attendance
19–21 April 1935	Nuclear Physics	35
27–29 April 1936	Molecular Physics	60
15–20 February 1937	Elementary Particles	26
21–23 March 1938	Stellar Energy	34
26–28 January 1939	Low Temperature	53
21–23 March 1940	Interior of the Earth	56
22–24 May 1941	Elementary Particles	33
23–25 April 1942	Stellar Evolution and Cosmology	25

*The data come from material presented by Karl Hufbauer at the 1980 meeting of the History of Science Society. The attendance figures include invited members and those present informally.

he fashioned the tools to solve the problem. But perhaps because of his fascination with problems of origins and genesis, he came to regard nucleosynthesis as the all-important problem and the explanation of the relative abundances of the elements the criterion by which the theory would be tested. He was unable to see that energy generation and nucleosynthesis need not be addressed simultaneously.

Bethe — always a theoretician who based his work on firm empirical data and sound phenomenological knowledge — decoupled the two aspects of the problem. Thus, after he had attended the 1938 Washington Conference and had been made aware of the problem, of the data, and of the tools at hand, Bethe was able to give the definitive answer to the problem of the energy generation in stars. He was awarded the Nobel Prize in 1967 for this work.

When World War II broke out in September 1939, Bethe certainly felt at home in the US. He had earned the affection and admiration of his colleagues at Cornell, had been recognized internationally as one of the outstanding theorists of his generation, and had married the woman he had fallen in love with. Despite the upheaval that Hitler's rise to power had engendered, for Bethe the decade had indeed been the "happy thirties." His own perspective on what had happened to him was movingly conveyed to Sommerfeld after the war, when Bethe was offered the chair in theoretical physics in Munich (see the box on page 143).

The 1930's in Retrospect

One can't help but be overwhelmed when looking back on Bethe's scientific output during the 1930's. More than half of the papers that were particularly meaningful to him and included in his selected works[9] were from that decade. Together with Wolfgang Pauli, Sommerfeld, Felix Bloch, Rudolf Peierls, Lev Landau, John Clark Slater, and Alan Wilson, Bethe was one of the founding fathers of solid-state theory (see *Physics Today*, June 2004, page 53 for David Mermin's interview with Bethe on solid-state theory).[10] He was one of the first theorists to apply group-theoretical methods to quantum mechanical calculations.[11] His theory of energy loss of charged particles in their passage through matter became the basis for extracting quantitative data from cloud chamber tracks and, later, nuclear emulsions.[12] After hole theory was formulated, his calculations of cross sections for pair production and bremsstrahlung became classics.[13] With Peierls he laid the foundations for understanding the structure of the deuteron, neutron–proton and proton–proton scattering, and the photodisintegration of the deuteron.[14] The Bethe Bible summarized what was known and understood in nuclear structure and nuclear reactions. And his paper on energy generation in stars solved that problem and created the field of nuclear astrophysics.

Along the way Bethe created little gems that proved seminal. In 1931, when he had decided "to treat the problem of ferromagnetism decently [by] ... really calculating the eigenfunctions," as he wrote to Sommerfeld, he first considered a one-dimensional chain of spins with an exchange interaction between nearest neighbors that was either positive, as in the Heisenberg model, or negative, as in the "normal" case. With the help of his famous ansatz — which in recent decades has found numerous other applications — Bethe started with the fully aligned ground state and determined the wavefunctions of states having an arbitrary number of reversed spins. Similarly, his refinement of the Bragg–Williams method[15] offered important insights into long-range correlations near the phase-transition point in alloys — and thus into phase transitions in general.

One can identify three fairly well delineated periods in Bethe's life through the mid-1950's. Until the early 1930's, it was German culture and German institutions that molded him. The two *Handbuch der Physik* articles are the fruition of stage one.

The period from the early 1930's till 1940 reflects his interactions with Fermi and with the physicists at Cambridge, Manchester, Bristol, and Cornell. It is also indicative of the sense of belonging these communities had offered him. Unlike the *Handbuch* articles, the Bethe Bible was undertaken on his initiative. It was designed to give to the American nuclear-physics community the theoretical perspectives that would direct their researches. The *Reviews of Modern Physics* articles and his solution of the problem of energy generation in stars[16] epitomize the capacities of the mature scientist who helped shape the new field of nuclear physics.

The third period, which began with the outbreak of World War II, saw Bethe solve, again on his own initiative, important problems in armor penetration and the physics and chemistry of shock waves. After Pearl Harbor, he acquired new authority at the Radiation Laboratory at MIT and at Los Alamos Laboratory: He became the charismatic leader of important divisions of those laboratories. The postwar years from 1946 to 1955 constituted one of the most exhilarating phases of Bethe's life, both scientifically and professionally. The stage for his activities became national and international. Bethe was at the center of important new developments in quantum electrodynamics and meson theory. He helped Cornell become one of the outstanding universities in the world. He was a much sought-after and highly valued consultant to the private industries trying to develop atomic energy for peaceful purposes. He was deeply involved and exerted great influence in issues concerning national security. He was happily married and the proud father of two very bright children. But the demands from his activities outside Cornell were enormous, the pace was grueling, and the activities were exacting a heavy toll both at home and in his research. In 1955 Bethe went to Cambridge University to spend a sabbatical year there. It was a year of taking stock and of narrowing his scientific focus.

Epilogue

In an article entitled "We Refugees," Hannah Arendt describes her experiences as a refugee first in France and then in the US following Hitler's rise to power:

> We lost our homes, which means the familiarity of daily life. We lost our occupation, which means the confidence that we are of some use in the world. We lost our language, which means the naturalness of reactions, the simplicity of gestures, the unaffected expression of feelings.[17]

Bethe's experience was almost the opposite of Arendt's. He did not lose the familiarity of daily life; on the contrary he became less isolated, and life in general became more intense, more rewarding, and more fulfilling for him. Nor did he lose his occupation; in fact, he obtained a temporary position that quickly became permanent and that allowed him to grow and to meet and surmount new challenges on a time scale much shorter than

Hans Bethe to Arnold Sommerfeld[5]

20 May 1947

I am very gratified and very honored that you have thought of me as your successor. If everything since 1933 could be undone, I would be very happy to accept this offer. It would be lovely to return to the place where I learned physics from you, and learned to solve problems carefully. And where subsequently as your Assistent and as *Privatdozent* I had perhaps the most fruitful period of my life as a scientist. It would be lovely to try to continue your work and to teach the Munich students in the same sense as you have always done: With you one was certain to always hear of the latest developments in physics, and simultaneously learn mathematical exactness, which so many theoretical physicists neglect today.

Unfortunately it is not possible to extinguish the last 14 years .... For us who were expelled from our positions in Germany, it is not possible to forget.

Perhaps still more important than my negative memories of Germany, is my positive attitude toward America. It occurs to me (already since many years ago) that I am much more at home in America than I ever was in Germany. As if I was born in Germany only by mistake, and only came to my true homeland at 28. Americans (nearly all of them) are friendly, not stiff or reserved, nor have a brusque attitude as most Germans do. It is natural here to approach all other people in a friendly way. Professors and students relate in a comradely way without any artificially erected barrier. Scientific research is mostly cooperative, and one does not see competitive envy between researchers anywhere. Politically most professors and students are liberal and reflect about the world outside — that was a revelation to me, because in Germany it was customary to be reactionary (long before the Nazis) and to parrot the slogans of the German National ["Deutschnationaler"] party. In brief, I find it far more congenial to live with Americans than with my German "Volksgenossen." [This word is identified with Nazi rhetoric, so there is a touch of sarcasm in Bethe using it. It might be rendered in English as "national comrade."]

On top of that America has treated me very well. I came here under circumstances which did not permit me to be very choosy. In a very short time I had a full professorship, probably more quickly than I would have gotten it in Germany if Hitler had not come. Although a fairly recent immigrant, I was allowed to work and have a prominent position in military laboratories. Now, after the war, Cornell has built a large new nuclear physics laboratory essentially "around me." And 2 or 3 of the best American universities have made me tempting offers.

I hardly need mention the material side, insofar as my own salary is concerned and also the equipment for the Institute. And I hope, dear Mr. Sommerfeld, that you will understand: Understand what I love in America and that I owe America much gratitude (disregarding the fact that I like it here). Understand, what shadows lie between myself and Germany. And most of all understand, that in spite of my "no" I am very grateful to you for thinking of me.

would have been the case had he remained in Germany. In addition, by virtue of the collective efforts he became engaged in, he became much more creative and productive. Until he had gone to England, Bethe was the sole author of his research publications. The prank with Beck and Riezler was his first collaborative effort, and a paper with Fermi his first true scientific collaboration. Many of Bethe's publications thereafter were joint efforts — with Peierls, with Livingston, with Bacher, with his graduate students and post-doctoral fellows.

Bethe didn't lose his language or suffer the consequences of its loss. He had secured his command of English during his first visit to England in 1930, and his stay in Manchester and Bristol had made him a native English speaker, except for a slight accent. Furthermore, the Anglo-American context evidently had allowed him to give genuine expression to his feelings. At a symposium on nuclear physics during the 1930's, Bethe entitled his talk "The Happy Thirties." Although he was referring primarily to developments in nuclear physics, his own personal and professional life had likewise been transformed for the better — despite the fact that "politically the thirties were anything but happy."[18]

References

1. Quotations without further references are taken from the interviews of H. A. Bethe by S. S. Schweber, summers 1992–1996.
2. M. Eckert, *Phys. Perspective* **1**, 238 (1999).
3. L. Hoddeson, G. Baym and M. Eckert, in *Out of the Crystal Maze: Chapters from the History of Solid State Physics*, eds. L. Hoddeson, E. Braun, J. Teichmann and S. Weart, Oxford Univ. Press, New York, 1992, p. 111, note 139.
4. H. A. Bethe, *Selected Works of H. A. Bethe: With Commentary*, World Scientific, 1997, p. 185.
5. M. Eckert and K. Märker (eds.), *Arnold Sommerfeld: Wissenschaftlicher Briefwechsel, Band 2: 1919–1951*, GNT-Verlag/Deutsches Museum, Berlin, 2003.
6. A. Sommerfeld and H. A. Bethe, *Handbuch der Physik* **24**, vol. 1 (1933); H. A. Bethe, *Handbuch der Physik* **24**, vol. 2 (1933).
7. H. A. Bethe and R. F. Bacher, *Rev. Mod. Phys.* **8**, 82 (1936); H. A. Bethe, *Rev. Mod. Phys.* **9**, 69 (1937); M. S. Livingston and H. A. Bethe, *Rev. Mod. Phys.* **9**, 245 (1937). Republished as H. A. Bethe, R. F. Bacher and M. S. Livingston, *Basic Bethe: Seminal Articles on Nuclear Physics, 1936–1937*, AIP Press, New York, 1986.

8. Livingston Farrand papers, #3-5-7, Division of Rare and Manuscript Collections, Cornell University Library.
9. H. A. Bethe, *Selected Works of H. A. Bethe: With Commentary*, World Scientific, 1997. The volume contains a bibliography of 290 of Bethe's publications (1928–1996) and reproduces 28 of his papers with brief introductory comments by Bethe.
10. A video recording, made 25 February 2003, of the conversation between Bethe and David Mermin on solid-state physics is available from the Cornell University Library at http://ifup.cit.cornell.edu/bethe.
11. H. A. Bethe, *Ann. Phys. (Leipzig)* **3**, 133 (1929). An English translation was published by Consultants Bureau, New York (1950), and is reproduced in Ref. 9, p. 1.
12. H. A. Bethe, *Ann. Phys. (Leipzig)* **5**, 325 (1930), translated and reproduced in Ref. 9, p. 767.
13. H. A. Bethe and W. Heitler, *Proc. Roy. Soc. London A* **146**, 83 (1934), reprinted in Ref. 9, p. 187.
14. H. A. Bethe and R. Peierls, *Proc. Roy. Soc. London A* **148**, 146 (1935), reprinted in Ref. 9, p. 223; H. A. Bethe and R. Peierls, *Proc. Roy. Soc. London A* **149**, 176 (1935), reprinted in Ref. 9, p. 235.
15. H. A. Bethe, *Proc. Roy. Soc. London A* **150**, 552 (1935), reprinted in Ref. 9, p. 245.
16. H. A. Bethe and C. L. Critchfield, *Phys. Rev.* **54**, 248 (1938), reprinted in Ref. 9, p. 347; H. A. Bethe, *Phys. Rev.* **55**, 434 (1939), reprinted in Ref. 9, p. 355; R. E. Marshak and H. A. Bethe, *Astrophys. J.* **91**, 239 (1940); H. A. Bethe, *Astrophys. J.* **92**, 118 (1940).
17. H. Arendt, in *The Jew as Pariah: Jewish Identity and Politics in the Modern World*, ed. R. H. Feldman, Random House, New York, 1978, quoted by K. Bell in *Displaced Persons: Conditions of Exile in European Culture*, ed. S. Ouditt, Ashgate, Burlington, VT, 2000.
18. H. A. Bethe, in *Nuclear Physics in Retrospect: Proceedings of a Symposium on the 1930's, May 1977*, ed. R. H. Stuewer, Univ. Minnesota Press, Minneapolis, 1979, p. 9.

Stellar Energy Generation and Solar Neutrinos*

John N. Bahcall and Edwin E. Salpeter

How energy is produced in the Sun and other stars was a big puzzle when Hans Bethe was a teenager in the early 1920's. When Arthur Eddington wrote his long *Encyclopedia Britannica* article on stars in 1911, he could report only on energy release from gravitational contraction, a clearly inadequate source. By 1920, however, Francis Aston had shown experimentally that the mass of the helium atom is slightly less than that of four hydrogen atoms.

Modern measurements tell us that the alpha particle's mass is less than four times the proton mass by 26 MeV, about 7%. Soon after Aston's discovery, Eddington suggested that the Sun's energy source might be the conversion of hydrogen to helium. If that were so, he argued, the Sun could shine for a very long time.

But no one in the 1920's knew how this putative fusion process might work. In the following decade, however, quantum mechanics became well established and nuclear physics was greatly stimulated by the discoveries of the neutron and the positron. Much of the progress on nuclei was due to Hans. His contributions included several papers with his friend Rudolf Peierls on neutron–proton scattering and on the deuteron. George Gamow's elucidation of Coulomb-barrier penetration was a huge step. But perhaps the greatest advance during that period was "the Bethe Bible," a set of three full issues (one each with Robert Bacher and Stanley Livingston as coauthor, and one alone) of *Reviews of Modern Physics* in 1936 and 1937.

---

*Reprinted with permission from *Physics Today*, October 2005, pp. 44–47. © American Institute of Physics.

The three review articles summarized all that was known, or even surmised, about nuclear physics at the time.

How the Sun Shines

Following up on a suggestion by Carl von Weizsäcker, Hans and Charles Critchfield, a graduate student of Gamow's, teamed up to calculate the rate at which two protons would fuse to form a deuteron with the emission of a positron and a neutrino.[1] They calculated a moderately accurate energy production rate for the p–p fusion chain as a function of temperature, but they did not have a detailed model of the solar interior from which to deduce realistic temperatures. Consequently, their inferred energy-production rate disagreed badly with the Sun's known luminosity. So Hans considered that work with Critchfield at the beginning of 1938 to be just an exercise rather than the start of a new science.

But his gloom vanished promptly after he attended a conference in Washington, organized by Gamow in March 1938. There he heard of new estimates of solar interior temperatures that brought his calculations into encouraging agreement with the Sun's luminosity. Exploiting his intimate knowledge of nuclear physics, Hans examined all reactions that might lead from hydrogen to helium. Studying all the exothermic reactions between a proton and the various isotopes of carbon and nitrogen, he found a cyclic phenomenon: Starting with ^{12}C, you don't simply end up with ^{16}O. Mostly, you come back to ^{12}C plus the desired helium nucleus.

Hans was able to calculate not only the reaction rates for this "carbon-nitrogen-oxygen cycle" as a function of temperature, but also the abundance ratios for the various isotopes that served as catalytic intermediates in the cycle. He correctly deduced that the CNO cycle and the p–p chain would supply about equal amounts of energy at a temperature of about 16 million kelvin. With reactants having high Coulomb barriers produced by nuclear charges of six and seven, the CNO thermonuclear reactions were more temperature-sensitive than the p–p chain.

It was already known that the luminosity of main-sequence stars much more massive than the Sun is a steeper function of temperature than it is for less massive stars. Working out the details of the CNO cycle took Hans only two weeks. It was immediately clear to him that this conjectured chain of fusion and breakup reactions must play an important astrophysical role. He concluded that stars significantly heavier than the Sun would shine via the CNO cycle and that lighter stars would shine via fusion initiated by the p–p reaction.

The overly simplistic models of the solar interior in use before the 1938 Washington conference gave central temperatures that were too high. Unlike those very inaccurate estimates, the central temperature in the Bethe–Critchfield paper was high by only 20%. That overestimate didn't make much difference for p–p energy production, but it did for the temperature-sensitive CNO cycle. So Hans inferred, incorrectly, that the Sun shines primarily via the CNO reactions. But his result for more massive main-sequence stars fit the observations remarkably well. As for stars that have left the main sequence, he already knew that red giants were altogether different.

Characteristically, until the end of his life, Hans remained interested in the question of the relative importance of p–p and CNO energy generation in the Sun. At age 96, he sent a short handwritten note to one of us (Bahcall), commenting on, and offering congratulations for, the exploitation of solar-neutrino observations to set a 7% upper limit on the CNO contribution to solar energy generation (see *Physics Today*, July 2002, page 13).

Hans's great 1939 paper was a landmark achievement that showed how stars shine.[2] It set the agenda for nuclear astrophysics for the next half century. Hans was awarded a New York Academy of Sciences prize for that work even before its publication. But his Nobel Prize did not come until 1967. The delay may have been due, in part, to the difficulty of deciding which of his many important contributions to recognize.

Two interesting aspects for the 1939 paper are not well known. First, given his important role in solar-neutrino studies in the 1980's and 1990's, it is amusing that Hans did not include a neutrino in any of the paper's nuclear reaction equations. For example, he left out the emerging neutrino in

$$p + p \rightarrow {}^2H + e^+ + \nu.$$

The explanation is simple. In 1934, Hans and Peierls, invoking dimensional arguments, had set an upper limit of 10^{-44} cm^2 on neutrino absorption cross sections.[3] Not unreasonably, they concluded that it would be "impossible to observe processes of this kind with the neutrinos created in nuclear transformation." The following year, Hans concluded from the absence of observed ionization by neutrinos that any neutrino magnetic moment must be much less than that of the electron.[4]

Second, Hans discussed the reactions ^{2}H + ^{2}H and ^{4}He + ^{4}He in the 1939 paper, but not the analogous reaction ^{3}He + ^{3}He. The ^{3}He reaction is, in fact, the dominant way of completing hydrogen burning to helium in modern solar models. So we asked Hans why he hadn't considered it in his comprehensive survey of nuclear fusion in main-sequence stars. Characteristically, he answered immediately: "I didn't think of it." Hans didn't waste time tooting his own horn.

Hans's 1939 paper laid the conceptual foundation for solving the energy-production problem in main-sequence stars. He explained, in outline, how the Sun shines. But he always wanted to make the theory more quantitative. Hans brought a few things up to date on the CNO cycle in a 1940 paper, but there was still a fair bit of unfinished business on both the p–p chain and the CNO cycle. Then, with the coming of war, astrophysics had to wait.

After the War

After his work directing the theory group at Los Alamos during the war, Hans returned to Cornell in 1945. In his reminiscences, Hans said modestly that he did not return to astrophysics in earnest until 1978, when he and Gerald Brown started to work on supernovae. This was indeed a major effort, which Gerry describes in his article on page 175. But Hans also had a tremendous, although indirect, impact on astrophysics in the 1950's and 1960's, partly by energizing William Fowler's group at Caltech. That stimulation involved many youngsters, including the two of us, who worked on subtopics of stellar energy generation — including the unfinished business of the p–p chain and the CNO cycle.

One example of Hans's prowess in making things clear-cut and simple was the effective-range theory[5] of nuclear scattering he invented in 1949. This new analysis technique soon enabled others to put the p–p reaction rate on a more secure footing. Hans's encouragement went beyond proton burning in main-sequence stars to another topic in which he had first become interested at the 1938 Washington meeting — that is, energy generation and the production of heavier elements in evolved stars consisting mainly of helium.

Besides his involvement in Fowler's endeavors, Hans had other indirect influences on astrophysics. Already in the 1940's, together with Robert Marshak, he had written some definitive papers on white dwarf stars. That work stimulated interest in transitional evolutionary stages preceding the white-dwarf stage, especially in central stars of planetary nebulae, where energy losses to neutrinos are important. Papers on the relevant neutrino reactions were written in the 1960's by Giles Beaudet, Hong-yi Chiu, Vahe Petrosian, and Hubert Reeves, all of whom had been Cornell graduate students at one stage or another.

Hans Bethe explaining the role of recently discovered vector mesons in nuclear forces to a press conference at the January 1967 meeting of the American Physical Society in New York. Seated facing the camera are (left to right) Rudolf Peierls, William Fowler, and Allan Bromley. Bethe worked with Peierls in the 1930's, and he inspired much of the later work of Fowler's astrophysics group at Caltech. (Courtesy of AIP Emilio Segrè Visual Archives.)

The Missing Solar Neutrinos

Starting in the late 1960's, Raymond Davis and coworkers reported that their neutrino detector, deep inside a South Dakota gold mine, recorded solar neutrinos at only about 1/3 the rate predicted by a rather precise solar model from which one would have expected better. Davis's radiochemical detector, a large vat of chlorine-rich fluid, was sensitive only to electron neutrinos — the only neutrino type produced by p–p or CNO reactions in the solar core.

Addressing the mystery of the missing solar neutrinos in 1986, the 80-year-old Hans wrote an influential paper that explained the Mikheyev-Smirnov-Wolfenstein (MSW) effect in language with which nuclear, atomic, and molecular physicists would be familiar.[6] In a Soviet journal, Stanislav

Raymond Davis and John Bahcall in 1966 at Davis's solar-neutrino detector deep inside the Homestake mine in South Dakota, shortly before the start of the experiment that first revealed the shortfall of neutrinos from the Sun. (Courtesy of the Institute for Advanced Study.)

Mikheyev and Alexei Smirnov had recently made the important suggestion that if neutrinos oscillate between different types, as Lincoln Wolfenstein and others had suggested, the metamorphosis of solar neutrinos might be resonantly amplified by matter effects in the Sun.

Using Wolfenstein's formalism, Mikheyev and Smirnov showed that the resonant neutrino conversion by matter outside the solar core could elegantly explain the discrepancy between Davis's observations and the solar model. The MSW effect would convert a large fraction of the electron neutrinos produced in the solar core to neutrino types invisible to Davis's detector.

Hans's explanation of the MSW effect — in terms of avoided crossings of nearly degenerate energy levels — appeared before the Mikheyev–Smirnov paper was published in English translation. The fact that he considered the MSW idea in particular, and new neutrino physics in general, worth exploring as a possible solution of the long-standing astrophysical problem energized a number of nuclear and particle physicists. Some of them went on to do extraordinarily beautiful and important solar-neutrino experiments. It suddenly became more fashionable to discuss ways in which physics beyond the standard model of particle theory could affect neutrino propagation from the interior of the Sun to detectors on Earth.

In 1990 Hans and one of us (Bahcall) demonstrated that if one compared the Davis results with newer data from a water-Cherenkov detector that had some limited sensitivity to other neutrino types, one had to conclude that the overall observational picture of the solar-neutrino deficit "requires new physics."[7] The case was made even clearer in 2001 by first results from the Sudbury Neutrino Observatory (SNO) in Ontario, whose heavy-water core could record neutrinos of all three types with equal sensitivity. Therefore SNO could confront the solar model in detail, irrespective of neutrino metamorphoses on the journey from the solar core to the terrestrial detector. The SNO experiment confirmed that the Sun's total output of neutrinos of all types was in excellent agreement with the solar-model prediction for the production of electron neutrinos in the Sun's core.

After the beautiful SNO result, Hans never wavered from his conviction, expressed in the 1990 paper, that physics beyond the standard model of particle theory was required to solve the solar neutrino problem. The standard model assumes, for simplicity, that all neutrino types are massless and therefore cannot metamorphose into one another. In 1993, with new gallium radiochemical detectors also reporting solar-neutrino shortfalls, Hans undertook a detailed Monte Carlo simulation of solar-model uncertainties.[8] The simulation showed unambiguously that without neutrino oscillation no plausible tweaking of the solar model was consistent with all the solar-neutrino data.

Before 1996, Hans often expressed the hope that he would learn the result of the SNO experiment in time for his 90th birthday. In fact, the results did not come until June 2001, when Hans was almost 95. Arthur McDonald, leader of the SNO effort, phoned Hans a few days before the public announcement to tell him that — although he couldn't reveal the result yet — he knew that we would be pleased to see the data. The result, when it was posted on the Web, was a strong confirmation of the solar model that had its beginnings with Hans's 1939 paper.

Collaborating with Hans was an honor and an enormous pleasure for both of us. He was a wonderfully enthusiastic coworker, with tremendous insight and mastery of an extraordinary range of physics. He was particularly skilled at making effective approximations.

We admired Hans as much for his personal qualities of decency, friendliness, honesty, and dedication to moral principles as for his greatness as a scientist. He used to advise his protégés: "Never work on a problem for

Bethe and his wife Rose in 1995.

which you do not have an unfair advantage." That was, in general, good advice. But for Hans himself, it was hardly a limitation.

> Those of us who had the good fortune of working with John Bahcall were particularly saddened by his sudden death shortly after he and I had finished this article. I always enjoyed writing papers with John. He brought refreshing originality to the beginning of the work, but then always ended up with meticulous quantitative detail. I will miss John, as will all the youngsters whom he had inspired. — E. E. S.

References

1. H. A. Bethe and C. L. Critchfield, *Phys. Rev.* **54**, 248 (1938).
2. H. A. Bethe, *Phys. Rev.* **55**, 434 (1939).
3. H. A. Bethe and R. Peierls, *Nature* **133**, 532 (1934).

4. H. A. Bethe, *Proc. Cambridge Philos. Soc.* **31**, 108 (1935).
5. H. A. Bethe, *Phys. Rev.* **76**, 38 (1949).
6. H. A. Bethe, *Phys. Rev. Lett.* **56**, 1305 (1986).
7. J. N. Bahcall and H. A. Bethe, *Phys. Rev. Lett.* **65**, 2233 (1990).
8. J. N. Bahcall and H. A. Bethe, *Phys. Rev. D* **47**, 1298 (1993).

Hans Bethe and Quantum Electrodynamics*

Freeman Dyson

From 2 to 4 June 1947, a carefully selected group of distinguished physicists assembled at Shelter Island, a small and secluded spot near the eastern tip of Long Island, to discuss the outstanding problems of physics. This was the first serious meeting of physicists who had played leading roles in World War II and then returned to the pursuit of peaceful science. The Shelter Island Conference succeeded in its purpose: It set the direction for physics for the next 30 years.

The main subject of discussion was the experiment of Willis Lamb and Robert Retherford, who used the tools of microwave spectroscopy, developed during the war for military purposes, to measure the fine structure of the energy levels of the hydrogen atom. The results showed a clear deviation of the observed levels from the predictions of the Dirac theory of the hydrogen atom. Lamb and Retherford measured a quantity that became known as the Lamb shift — the frequency of a microwave field that induced transitions between the lowest two excited states of the hydrogen atom. According to the Dirac theory, the two states should have had equal energy and the Lamb shift should have been zero. Lamb measured it to be 1000 megahertz, with an uncertainty of a few percent. The discrepancy was far outside the limits of possible experimental error.

Many people at the conference, including Victor Weisskopf and Robert Oppenheimer, suggested that the deviation resulted from quantum fluctuations of the electromagnetic field acting on the electron in the atom. Such fluctuations would give the electron an additional energy, called the self-energy. It was well known that the existing theory of quantum electro-

---

*Reprinted with permission from *Physics Today*, October 2005, pp. 48–50. © American Institute of Physics.

dynamics (QED) gave an infinite value for the self-energy and was therefore useless. Physics had reached an impasse. On the one hand, the Lamb experiment gave clear evidence that the effects of electromagnetic quantum fluctuations were real and finite. On the other hand, the existing theory of QED gave infinite and absurd results. It was obvious to everyone at the meeting that breaking the impasse would require a new idea.

Hendrik Kramers, one of the few non-US attendees, provided the new idea, which was named "renormalization." That name was already familiar to physicists in 1947; it had been used in a similar context by Robert Serber in 1936. Kramers had come from the Netherlands to spend a term as a visiting scientist at the Institute for Advanced Study in Princeton, New Jersey. He remarked that the observed energy of an electron according to QED is the sum of two unobservable quantities: a bare energy, which is the energy that an electron is supposed to have when it is uncoupled from electromagnetic fields, and the self-energy, which results from the electromagnetic coupling. The bare energy appears in the equations of the theory but is physically meaningless, since the electromagnetic coupling cannot really be switched off. Only the observed energy is physically meaningful. The point of renormalization was to get rid of bare energies and replace them with observed energies.

Kramers proposed that the results of the Lamb experiment should be calculated in terms of observed energies, with all mention of bare energies removed. He conjectured that when the bare energies were eliminated from the calculation, the infinite self-energies would cancel out and the calculated value of the energy difference that Lamb and Retherford measured would become finite. Kramers sketched a simple model of an electron for which the calculation could be done and the result was finite. But he did not know how to carry through the calculation for a real electron in a real hydrogen atom. Nobody at the meeting knew how to do a realistic calculation following Kramers's idea. Except Hans Bethe.

After the meeting, Bethe traveled by train from New York to Schenectady, a distance of 75 miles. On the train he finished a calculation of the Lamb shift for a real electron. The value that he found was 1040 megahertz, a result agreeing pretty well with Lamb's experiment. On 9 June he wrote a paper summarizing his calculation,[1] and sent it to the other participants of the Shelter Island meeting. The paper, two pages long with only 12 equations, was received by the *Physical Review* on 27 June and published on 15 August. It was a turning point in the history of physics. Before it appeared, the prevailing view of such experts as Niels Bohr and Oppenheimer

Richard Feynman explains a point during the 1947 Shelter Island Conference. Standing (from left to right) are Willis Lamb, Karl Darrow, Victor Weisskopf, George Uhlenbeck, Robert Marshak, Julian Schwinger, and David Bohm. Seated are Robert Oppenheimer, Abraham Pais, Feynman, and Herman Feshbach. (Courtesy of the National Academy of Sciences.)

was that existing theories of particle physics were fundamentally flawed and that progress could come only from revolutionary new concepts. After the paper appeared, the experts began to think that the existing theory of QED was physically correct and needed only some new technical tricks to make it mathematically consistent and practically useful.

How did it happen that Hans Bethe was the one who broke the impasse? There were two reasons. First, Bethe understood that the reaction of the electron to electromagnetic quantum fluctuations was mainly a nonrelativistic process and could be calculated using ordinary nonrelativistic quantum mechanics. Everyone else at the meeting assumed that a calculation of the Lamb shift had to use the notoriously difficult and complicated theoretical apparatus of relativistic QED. Only Bethe had the courage to plunge ahead with a calculation using old-fashioned nonrelativistic quantum mechanics to describe the hydrogen atom. Ignoring relativity made the calculation enormously simpler.

Second, Bethe was uniquely prepared by his previous training and experience to do the calculation and get the right answer. By a happy combination of circumstances, he had unrivaled knowledge of QED, of atomic physics, and of the art of calculating physical processes in which QED and atomic physics came together. It was his habit, when confronted with any physical problem, to sit down and calculate the answer. So, when the problem of the Lamb shift arose, it was natural for him to be the one who sat down and calculated the answer. To explain how Bethe acquired the knowledge and skills that enabled him to calculate the answer, I go back 20 years.

Mastering QED

The history of QED began at the end of the 1920's with three independent lines of development. First, in 1927 Paul Dirac developed a version of QED in which both electrons and photons are treated as particles. Second, in 1929 Werner Heisenberg and Wolfgang Pauli used the complicated mathematics of quantum field theory to develop a version in which both electrons and photons are treated as fields. Third, in 1930 Enrico Fermi developed a version in which the electrons are treated as particles and the photons as fields. Fermi's version was typical of his work — mathematically simple and physically transparent. Fermi intended it for use, not for ornament. Of the three versions of QED, Fermi's was the most practical and the best suited for solving real problems. Bethe had the good sense to go to Rome in 1931 and learn QED from Fermi. The two men published a paper[2] using Fermi's version of QED to calculate the retarded interaction between two electrons caused by the emission and absorption of photons. This collaboration gave Bethe his first experience of using QED for practical purposes and provided the foundation for his later mastery of QED.

After Bethe returned to Germany from Rome in 1932, he wrote a monumental review article for the *Handbuch der Physik* on the quantum theory of one-electron and two-electron systems.[3] That review, concerned mainly with hydrogen and helium atoms, provided an outstanding display of Bethe's thoroughness and attention to details. All the wavefunctions of states of the hydrogen atom are precisely calculated in several different representations, using spherical, cylindrical, or parabolic coordinates as required for the interpretation of various experimental observations. After writing the review,

Bethe probably had a deeper understanding of the hydrogen atom than anyone else on Earth. He knew under which conditions a relativistic theory of the atom was necessary and under which conditions a nonrelativistic theory was sufficient.

After leaving Germany in 1933 and finding temporary refuge in Manchester, England, Bethe collaborated with his fellow refugee Walter Heitler, who was also expert in QED. Together they carried out the first precise relativistic calculation of the two dominant processes of high-energy electrodynamics — the production of photons by high-energy electrons and the production of electron–positron pairs by high-energy photons passing through matter.[4] The Bethe–Heitler calculation of photon production (called by its German name, "bremsstrahlung") and pair production was the most important achievement of QED in the 1930's. The calculation was a tour de force of analytical skill, and it explained quantitatively the observations of cosmic-ray showers, in which high-energy electrons passing through the atmosphere generated a multitude of lower-energy electrons, positrons, and photons traveling together in the same direction. Bethe and Heitler called attention to the fact that not all high-energy cosmic rays generated showers.

The theory of QED appeared to apply well to some cosmic rays and not to others. Bethe and Heitler thought that the absence of showers accompanying some of the fast particles indicated a breakdown of QED at high energies. Some years later the particles that did not produce showers were identified as mesons, and the evidence for a breakdown of QED disappeared. The Bethe–Heitler calculation remained the gold standard for careful and accurate work in high-energy physics.

In 1935 Bethe found his permanent home at Cornell University. For the rest of the 1930's, he was mainly occupied with nuclear physics and with understanding the nuclear reactions that keep the Sun and the stars shining (see the article by John Bahcall and Ed Salpeter on page 147). He received a richly deserved and long overdue Nobel Prize for this work in 1967. And still, in my opinion, Bethe's calculation of the Lamb shift was a more profound and, in the long run, a more important contribution to science. It broke through a thicket of skepticism and opened the way to the modern era of particle physics. It showed us all how to connect QED with the real world.

After the Lamb shift calculation, Bethe continued to work out the consequences of QED. The most original of his later discoveries was the Bethe–Salpeter equation, a relativistic wave equation for bound states of two particles.[5] With his student Leonard Maximon he worked out a new version of the Bethe–Heitler calculation of bremsstrahlung and pair production; they

Colleagues gather to discuss high-energy nuclear and particle physics at a 1955 conference in Rochester, New York. In front (from left to right) are Pierre Noyes, Freeman Dyson, Jack Steinberger, Richard Feynman, and Hans Bethe. (Courtesy of AIP Emilio Segrè Visual Archives, Marshak collection.)

used more modern methods and gave results that are more accurate when the atoms causing those processes are heavy.[6] With Ed Salpeter he published a greatly revised and expanded version of the *Handbuch der Physik* article on quantum mechanics of one-electron and two-electron systems.[7]

Hans Bethe never ceased to give help and encouragement to Toichiro Kinoshita and other colleagues at Cornell, who continued to push the calculation of QED processes to fourth, sixth, and eighth order of perturbation theory and thus keep pace as experiments in atomic physics became more and more accurate. Theory and experiment are now both accurate to 12 significant figures, and QED still stands confirmed. The confirmation of QED as the most precisely tested of all the laws of nature is one of the great triumphs of 20th-century science. We owe that triumph chiefly to the vision of two men, Willis Lamb and Hans Bethe.

References

1. H. A. Bethe, *Phys. Rev.* **72**, 339 (1947).
2. H. A. Bethe and E. Fermi, *Z. Phys.* **77**, 296 (1932).
3. H. A. Bethe, *Handbuch der Physik* **24**, 273 (1933).
4. H. A. Bethe and W. Heitler, *Proc. Roy. Soc. London A* **146**, 83 (1934).
5. H. A. Bethe and E. E. Salpeter, *Phys. Rev.* **84**, 1232 (1951).
6. H. A. Bethe and L. C. Maximon, *Phys. Rev.* **93**, 768 (1954).
7. H. A. Bethe and E. E. Salpeter, *Handbuch der Physik* **35**, 88 (1957).

Hans Bethe and the Theory of Nuclear Matter*

John W. Negele

Prior to World War II, nuclear physics was a phenomenological science, and Hans Bethe was unrivaled in his comprehensive mastery of nuclear phenomena, experimental data, and descriptive models. As described in other articles in this special issue of *Physics Today*, by applying the emerging phenomenology Bethe achieved remarkable successes that ranged from understanding the energy production in stars to guiding the harnessing of nuclear fission as part of the Manhattan Project.

The post-war era offered Hans the opportunity to return to nuclear physics and approach the subject from a deeper theoretical perspective: Understanding the many-body structure and properties of nuclei directly in terms of the underlying nuclear interaction. He was freed from the applied-physics demands of the war effort and could again pursue theoretical physics for its own sake. His goals were to understand why the shell model worked in the presence of nuclear forces containing strongly repulsive short-range components; to understand nuclear collective motion; and to calculate the binding energies, excitation energies, nuclear charge distributions, and deformations of atomic nuclei. Once terrestrial nuclei were sufficiently understood, he could use that knowledge as a basis for studying the properties of dense matter in supernovae and neutron stars.

Foundations

One of the keys to his success in confronting complex problems was that

---

*Reprinted with permission from *Physics Today*, October 2005, pp. 58–61. © American Institute of Physics.

Hans could complement his extensive knowledge of a field with an acute ability to separate the essentials from the nonessentials. So it is illuminating to see what he regarded as the essential issues in nuclear physics, and how those guided his approach to the nuclear many-body problem.

I had an inside view of that approach when I took his nuclear-physics course in the fall of 1967 as his graduate research student at Cornell University. That same fall, Hans received the Nobel Prize in Physics. The announcement was made on a lecture day, so we students had the pleasure of watching his multitalented secretary, Velma Ray, tuck his tie neatly under his collar for the photographers. Then we got to hear Hans inform them politely but firmly that they needed to finish their task quickly because he had to start his lecture (see Fig. 1). Another fond memory connected with the prize was Hans's crash program for learning everything that had happened in stellar evolution since his 1939 paper on energy production in stars.[1] My reward for helping prepare the graphs for his Nobel lecture[2] was a detailed explanation of the physics each graph displayed.

Fig. 1 On the day he learned that he would receive the 1967 Nobel Prize in Physics, Hans Bethe insisted on teaching his usual nuclear-physics class. The day's lecture included some topics discussed in this article. Just below Hans's hand is an equation for the reaction matrix, and the graph at the extreme right shows both the probability depletion at small separation, or "wound," in the two-particle wavefunction and the "healing" of that wound at large separation. (Courtesy of Cornell University.)

As his starting point for nuclear physics, Hans chose the nuclear force, whose long-range behavior was given uniquely by a one-pion exchange potential. Before the war, Hans had worked out, in his own characteristic physical way, the spin-dependent part of that long-range potential, and he used it with an appropriate cutoff to calculate the properties of the deuteron.[3] The short-range potential could not be calculated by any known theory, so Hans thought about the most physical way to deal with the problem. The first step was to understand that very different potentials produce the same low-energy scattering behavior; that insight led to his famous simplified derivation of effective-range theory.[4] He then adopted the pragmatic approach of using physical arguments to parameterize the strong short-range repulsion, determining the parameters from scattering phase shifts, and then studying the many-body physics the complete potential produced.

An essential feature of nuclear physics, in Hans's view, was that nuclei behaved like quantum liquid drops. Moreover, the binding energy per particle and equilibrium density of the quantum liquid could be deduced from electron scattering experiments and the few-parameter semi-empirical mass formula derived from measurements of nuclear masses. Thus, the central task of nuclear many-body theory was to calculate, from the two-body potential, the properties of nuclear matter — an infinite system of neutrons and protons at equal density interacting through nuclear forces but without Coulomb interactions. Once nuclear matter was properly understood, one could apply many-body theory to finite nuclei and to the material of neutron stars.

Hole-line Expansion

The starting point of Hans's solution of the nuclear-matter problem was Keith Brueckner's pioneering approach to solving the two-body scattering problem in the nuclear medium. Brueckner rearranged perturbation theory so that the contribution to the total energy at each order was proportional to the number of particles. Thus, the energy per particle was manifestly finite even in the limit of nuclear matter. Hans, in his usual style, embarked on a systematic program to calculate properties of nuclear matter; his approach was based on a diagrammatic expansion of perturbation theory in a series ordered by the number of interacting particles, that is, the number of so-called hole lines in the contributing diagrams.

The first term in the series involved two particles and incorporated a sum of all two-body scattering processes in the background of other particles, as in Brueckner theory. Hans calculated the reaction matrix, which sums all two-body ladder diagrams. These correspond to processes in which two particles in the Fermi sea rescatter any number of times into unoccupied intermediate states and finally return to the original two states. He found it most physical to formulate his calculation in terms of the two-body wavefunction. As a consequence of the Pauli principle, there is no scattering phase shift. The two-body wavefunction thus has a "wound," where the probability density is depleted due to the short-range repulsive potential, that "heals" to the free, independent-particle wavefunction at large relative separations.[5]

Analytic solutions of special cases were a regular source of insight for Hans, so with Jeffrey Goldstone he solved the case of an infinite hard-core potential. Their solution became a touchstone for calculations with realistic potentials. With Baird Brandow and Albert Petschek, he introduced an approximation in which the particle energy spectrum was treated as quadratic in momentum. That approximation enabled them to convert the integral scattering equation into a differential equation that is easily solved, treat the corrections perturbatively, and include the full complexity of realistic nucleon–nucleon potentials at the two-body level.

Goldstone derived the linked cluster theorem using many-body diagrams,[6] and the resulting Goldstone diagrams became Hans's organizing language for the nuclear many-body problem. In contrast to Feynman diagrams, Goldstone diagrams retain a fundamental distinction between hole lines, which represent propagators for normally occupied single-particle states, and particle lines, which represent propagators for unoccupied states.

Hans showed that the Goldstone-diagram expansion for the binding energy of nuclear matter does not converge in powers of the reaction matrix. Rather, he demonstrated, one must rearrange the expansion in powers of the density or, equivalently, in the number of independent hole lines. Hence, as the next step in the hierarchy, he formulated what is now called the Bethe–Faddeev equation, which sums all three-body ladder diagrams. In this case, three particles in the Fermi sea pairwise scatter to unoccupied states any number of times via the reaction matrix before finally returning to the original three states. Hans's approach generalized to the nuclear medium Ludwig Faddeev's technique for three-body scattering in free space. Hans found it most physical to formulate the problem in terms of the three-body wavefunction, and developed the analytical tools to evaluate the three-body contributions to the binding energy.[7] Terms in the density expansion that correspond to greater numbers of hole lines are treated analogously.

Hans and coworkers fleshed out the basic ideas just sketched. Hans's student Roderick Reid constructed a potential that had a strongly repulsive core and fit experimental phase shifts. Former student Benjamin Day used that potential in an extensive set of nuclear-matter calculations. Overall, the results were quite impressive. Indeed, the hole-line expansion converged as expected — the three-hole-line contributions changed the total potential energy by approximately 13% and the four-hole-line contributions changed it by an estimated 3%. The binding energy per particle of 17 MeV was quite close to the mass-formula value of 16 MeV. The equilibrium density, however, was approximately 30% higher than the value inferred from electron scattering experiments on nuclei. Hans interpreted the convergence as validation of the nuclear-matter theory, and attributed the incorrect equilibrium density to the omission of explicit three-body forces.

Nuclei and Neutron Stars

Hans's view was that once nuclear matter was under control, finite nuclei and the matter in neutron stars would follow. In keeping with his love of tractable analytical approximations, his starting point for finite nuclei was the Thomas–Fermi approximation. In that scheme, each small region of the nucleus is treated as if it contained nuclear matter with the same density. Hans's theory gave a qualitative description of the surface energy and surface thickness of large nuclei.

After pulsars were observed and subsequently interpreted as neutron stars, Hans applied nuclear-matter ideas to the charge-neutral matter in those stars. With Gordon Baym and Chris Pethick, he considered the range of configurations that result as the density in a star increases from low densities characterized by well-separated nuclei up to densities typically found inside nuclei.[8] The new feature is that as the density increases, the Fermi energy of the electrons increases, and so it becomes energetically favorable for an electron and proton to combine to make a neutron and also a neutrino that escapes from the star.

Using results for nuclear matter at unequal neutron and proton densities, the researchers were able to calculate the equation of state of matter as neutrons begin to "drip" out of nuclei and form a low-density neutron gas between them. Hans and collaborators could then follow the increase of the neutron density and calculate the eventual merging of individual nuclei into a uniform gas with a high density of neutrons, a low density of protons, and an equally low density of electrons.

With Mikkel Johnson, Hans used nuclear-matter theory to calculate the equation of state of dense matter for several potentials similar to the Reid potential. The two then teamed with Robert Malone to work out the properties of neutron stars based on each equation of state. Their approach, based on phenomenological nuclear potentials fit to phase shifts, remains the most viable method for treating the equation of state for densities up to several times those occurring in nuclei.

Legacy

The body of work Hans contributed to nuclear physics spanned several decades and played an essential role in laying the foundation for an understanding of nuclear structure in terms of the underlying nuclear force. What physicists learned subsequently about strong interactions only reinforces the wisdom of Hans's approach. The nucleon is now understood to be a composite system composed of quarks and gluons, with a spatial size of about one fermi (10^{-15} m). Nucleons, in turn, combine to form a nucleus that typically has a size of several fermis. Because the scales of the nucleon and nucleus are not well separated, one cannot derive an unambiguous nuclear potential at short distance: The only option is to use the pion contribution to describe long-distance behavior and to use an effective theory derived from scattering properties for short-distance behavior.

Furthermore, modern local effective field theory necessarily gives rise to many-body forces whose parameters must be determined from properties of many-body systems. That result is consistent with the idea that a three-body force should be introduced to make the equilibrium density of nuclear matter agree with the value determined from finite nuclei. Indeed, physicists working with mean-field theories containing an effective interaction derived from nuclear matter and a three-body interaction that yields proper equilibrium densities have found them to be extremely successful in determining the binding energies, excitation energies, nuclear charge distributions, and deformations of nuclei throughout the periodic table.

Hans's approach to theoretical physics offers many valuable lessons. He refused to be stymied by a complex problem or incomplete information, and never hesitated to make a physical approximation, if necessary, so that he could proceed toward his objective. He was fearless in introducing a physical cutoff to avoid singularities — whether it be his truncation of the short-range one-pion potential in nuclear physics or his truncation of the high-frequency fluctuations of the electric field as part of his famous estimate of the Lamb shift, which he formulated before renormalization theory was developed for quantum electrodynamics. Analytical calculations, often coupled with simplifying approximations, were among his standard tools for obtaining insight into complicated problems. When, during my student days, I once expressed admiration for the way he reduced a problem to an analytically tractable form, he quipped "Yes, when you get old, you have to reuse the same old tricks."

His methodical approach to complicated problems like the many-body theory of nuclei was truly impressive. Starting from a clear vision of the essential issues and committing to the necessary approximations, he would map out a conceptually clear but calculationally complicated program, and then systematically plow through a huge number of steps.

One can learn much from his expositions. In his publications, he explained every aspect of a problem clearly and completely, without suppressing details. He carefully studied the work of others and cited it meticulously. Once, after he suggested that I include all the relevant details in a paper I was writing, I expressed concern about consequent page charges. He replied that part of the cost of research is publication. He was also admirably straightforward in discussing errors. In a review article about summing three-particle ladder diagrams, he started a paragraph with "Bethe originally proposed ..." and went on to say "This argument is wrong," followed by a gracious footnote crediting a discussion with colleague Thorolf Dahlblom.

Out of the Ordinary

Hans was an extraordinary adviser and mentor. When a student went to see him in his office, he would always interrupt a physics calculation to talk. During my years of graduate study, he traveled frequently to the national labs and the federal government in Washington, DC, to advise them on research, arms control, and the Nuclear Test Ban Treaty. So I was deeply impressed that he would come to Newman Laboratory on campus on Saturdays and make himself available to students. The flip side of that availability was that, as I learned early on, one should be prepared to receive a call from him at 5:00 p.m. on a Saturday afternoon to discuss a research idea.

In this age when it appears obligatory to expend so much professional time on university and national committees, fundraising, writing and reviewing proposals, and when the pendulum has swung so far toward promoting discussions, seminars, and interactions at every available opportunity, it is inspiring to recall Hans's enthusiasm for sitting quietly alone in his office focusing intensely on his work. That modus operandi served him well. Indeed, I vividly remember discussing a nuclear-physics problem with another of his graduate students in the coffee room. Hans overheard our discussion and said: "I don't see why you are talking about this problem when either of you is capable of sitting down and solving it." I also remember the pleasure he took in performing his legendary numerical calculations in his head; Hans could remember logarithms or expand functions as necessary and quote answers to several decimal places.

He and his wife Rose, seen together in Fig. 2, were always warm and hospitable to students and visitors. When Judit Nemeth arrived from Hungary as a postdoc, Hans personally met her at the airport and helped her get settled. Rose's dinner parties were wonderful affairs where colleagues, friends, students, and visitors were warmly welcomed and treated like family.

I also recall with gratitude Hans's response to a request I made in 1968, during the height of the Vietnam War. I had asked permission to interrupt

Fig. 2 Enjoying a moment at the 1967 Nobel Prize ceremonies in Stockholm are Hans Bethe and his wife Rose. (Courtesy of Cornell University.)

my graduate studies to go to New Hampshire with some other Cornell graduate physics students and work for an extended period on Eugene McCarthy's antiwar primary campaign. He responded that ordinarily a young scientist like me should devote himself exclusively to his work so as to have maximal influence later. But in this case, the war was so terrible that I should go with his blessing. Looking back, I find it noteworthy that in Concord, Nashua, and other New Hampshire towns where Cornell physicists labored to explain to nonacademic citizens why we believed the war was wrong, people voted solidly for McCarthy and that the New Hampshire primary marked the beginning of the end of the war. Hans's wisdom in balancing profession and patriotism is as relevant today as it was then.

In theoretical physics and in life, Hans continues to be a source of inspiration for all those whose lives he touched.

References

1. H. A. Bethe, *Phys. Rev.* **55**, 434 (1939).
2. H. A. Bethe, http://nobelprize.org/physics/laureates/1967/bethe-lecture.html.
3. H. A. Bethe, *Phys. Rev.* **57**, 260 (1940); **57**, 390 (1940).
4. H. A. Bethe, *Phys. Rev.* **76**, 38 (1949).
5. H. A. Bethe, *Phys. Rev.* **103**, 1353 (1956).
6. J. Goldstone, *Proc. R. Soc. London, Ser. A* **239**, 267 (1957).
7. See R. Rajaraman and H. A. Bethe, *Rev. Mod. Phys.* **39**, 745 (1967).
8. G. Baym, H. A. Bethe and C. J. Pethick, *Nucl. Phys.* **A175**, 225 (1971).

Hans Bethe and Astrophysical Theory*
Gerald E. Brown

I got to know Hans Bethe in the 1950's, when I was a lodger at the home of Rudolf and Eugenia Peierls in Birmingham, England. Bethe and Peierls had been close friends ever since they were graduate students together under Arnold Sommerfeld; the two went on to devise a comprehensive theory of the deuteron. Their comradeship affected many people, including some of the authors in this special issue of *Physics Today*. My room still had University of Canberra stationery from when Edwin Salpeter lived with the Peierlses, and I bought the bicycle Freeman Dyson had left.

During his 1955 sabbatical at the University of Cambridge, Hans worked on the nuclear many-body problem and, in particular, on difficulties associated with Brueckner theory. (See the article by John Negele on page 165.) Shortly after that, I began work on applying to finite nuclei the effective interactions he had obtained for infinite nuclear matter, and visited him often at Cornell University. After he retired from Cornell in 1976, he and I worked together on astrophysics problems; Hans also continued to work on neutrino physics, arms control, and energy issues.

I soon learned why Hans had no long-term collaborators earlier in his life, aside from Peierls. (The two are shown together in Fig. 1.) Even after he retired, it was nearly impossible to keep up with him. Not that he worked rapidly, but right away he could identify the essential physics and see the light at the end of the tunnel. Then, like a bulldozer, he moved toward that light, undeterred by temporary obstacles. The scope of problems he could solve pretty much had no limit: In that sense, I think he was the most powerful scientist of the 20th century.

---

*Reprinted with permission from *Physics Today*, October 2005, pp. 62–65. © American Institute of Physics.

Fig. 1 Rudolf Peierls (left) poses with Hans Bethe. Before Hans's nominal retirement from Cornell University in 1976, Peierls was his only long-term collaborator. This photograph of the two friends was taken in 1971.

He was famous for his statement "I can do that!" A couple of years after his retirement, I unleashed Hans's arsenal of physics knowledge and experience on a problem worthy of his abilities.

Not Enough Entropy

On the morning of 1 April 1978, my wife Betty and I picked up the Bethes from the airport in Copenhagen, then proceeded to take them to the house where they were renting some rooms. Hans had sprained his ankle on Mount

on in Turkey, and on the way I said that the accident was a bad omen
 his work on the pion–nucleon many-body problem, in which each new
 her-order term was about as large as all the previous lower-order terms
 together. Hans asked, "What should we work on?" I replied, "Let's
 k out the theory of supernovae." Hans countered that he didn't know
 thing about them.

I told him that workers in the field had the nuclear physics all wrong
and that I knew how to correct it. I'd written a research paper with James
Lattimer and Ted Mazurek showing that the many excited states of nuclei
had essentially been left out in works to date.[1] But, I told Hans, in order
to work out the theory, I needed an expert on explosions. Hans, who had
worked at the Los Alamos Laboratory as head of the theory group for the
Manhattan Project, admitted that he knew something about them. We
delivered Hans and his wife Rose to their apartment, and then I went to
work at the Niels Bohr Institute. Before going home that afternoon, I left
on Hans's desk a computer printout of a large star that Stan Woosley had
evolved up to the point of collapse. At that point, stellar fusion has converted
all the core material into iron, the nucleus with the greatest binding energy
per nucleon, and no additional energy can be produced via stellar burning.

I came in the next morning and went to see Hans. He said, "The entropy
in the iron core is very low, less than unity [in units of Boltzmann's constant
k_B] per nucleon."

"So what?" I asked.

He said, "That means that the iron core will collapse, without being held
up, until the iron nuclei merge into nuclear matter at nuclear-matter density.
There isn't enough entropy for them to break up." I said that since World
War II, all supernova calculations had shown that the collapse is held up at
about 1/1000 of nuclear-matter density.

"They're wrong!" he replied.

Hans was right. Stellar collapse is slow when considered on the time
scale of the strong interaction, so, setting the weak interactions aside for
the moment, entropy is conserved. And there simply isn't enough entropy
for iron to break up. The weak interactions actually decrease the entropy
in the core via neutrinos that carry energy and entropy out of the star. To
stop the infall of matter, a strong repulsive force is necessary, and such a
force is available only if the density is well above nuclear-matter density. In
that regime, the high compression modulus of nuclear matter can stop the
collapse.

Stellar Explosions

Because each iron nucleus in the core contributes about 1/27 as much to a star's pressure as do the relativistic, highly degenerate electrons of the atom, the collapse of the core is comparable to the collapse of a white dwarf. And as discovered by Subrahmanyan Chandrasekhar, the maximum dwarf mass M that can be supported by relativistic degenerate electron pressure is given by $M = 5.76 M_\odot Y^2$, where $M_\odot$ is the mass of the Sun and Y is the ratio of the number of electrons to the number of nucleons (protons and neutrons). In relativistic white dwarfs, $Y = 1/2$, but in large stars some electrons are captured by protons before collapse, so that Y is typically about 0.43 in the crucial stage of the collapse. Thus the mass of the so-called homologous core of a collapsed large star is about 1 solar mass, or perhaps a little greater. The result of the collapse, astrophysicists believe, is a neutron star, typically with a mass of $1.4 M_\odot$.

By 1978 the basics of the collapse were understood. The homologous core collapses subsonically, but the matter beyond the core's outer edge is supersonic. When the density at the center of the star exceeds the density of nuclear matter, pressure waves that travel at the speed of sound are formed. Those waves, which tend to bring infalling matter to rest, can only get as far as the edge of the homologous core: Were they to go beyond, they would be swept back in by the supersonic flow outside. Pressure and density discontinuities build up at the edge of the homologous core. Eventually they lead to a shock wave that blows off the part of the star well outside the homologous core. Explosions of stars that had previously converted all their core material into iron are called type II supernovae. They provide the universe with the heavy elements such as carbon and oxygen that make up human bodies.

The devil, of course, is in the details. Hans could solve almost any problem, but in attempting to understand the explosion, he probably had not encountered one more suited to his abilities and experience — especially in light of what he learned at Los Alamos. Alas, when the details were worked out, the better the calculation, the less well the explosion succeeded at blowing off the outer part of the star. The shock wave spent so much energy in dissociating the iron outside of the homologous core — about 10^{51} ergs (10^{44} joules) for each $0.1 M_\odot$ of iron dissociated — that it didn't have enough remaining energy to blow off the outer region; instead it just traveled some few hundred kilometers and stalled, until the investigator ran out of computer time.

The energy scale of 10^{51} ergs came up often enough in our work that I gave it a special name: the foe (short for 51 ergs). In 1990 Hans and Pierre Pizzochero[2] analyzed the speed of the hydrogen recombination wave in the supernova SN1987a and showed that the total mechanical energy in the explosion was 1–1.4 foe. Thus, dissociation of $0.1 M_\odot$ of iron would remove from the shock wave about as much energy as was contained in the mechanical part of the explosion.

But the explosion has another source of energy: gravitational binding. As the core of the large star settles into the tightly bound neutron star, neutrinos carry off about 300 foe of gravitational binding energy. So Hans tried to see if some of that energy could be deposited into the matter outside the homologous core but inside the shock. Then, through convection, the hot matter could move up behind the shock and repower it.

In late 1987 Hans discussed his convection ideas at the Nuclear Theory Institute in Seattle. When he came to the calculation, he said he had two coupled differential equations that he had to solve numerically. For Hans, "numerically" meant with his slide rule. Throughout our entire supernova work, he would take out sections of computer output, check different quantities against each other, and then pose questions to whoever had made the computation (see Fig. 2).

Fig. 2 The slide rule was Bethe's calculational tool of choice. In this photo, taken at Cornell University around 1986, Bethe analyzes computer output. (Photograph by Kurt Gottfried, courtesy of AIP Emilio Segrè Visual Archives.)

To this day, calculated explosions have yet to achieve success. Investigators are refining ideas about convection and relaxing assumptions about sphericity to get the explosions to work.

About the Future

For nearly 25 years, Hans and I spent every January together in California. We'd share a condominium in Santa Cruz or Santa Barbara, or, most often, in Caltech housing. I did the cooking since Rose was looking after her parents and my wife Betty was caring for our children. But in the last years, the four of us were all joyfully together (see Fig. 3). After supper Hans and I would discuss history, politics, and so forth, always in a structured way. At times, he would lapse into German and reminisce about his childhood and youth. He greatly appreciated my ability in German and gave me Johann Wolfgang von Goethe's book of poems *Das Leben, es ist gut* for a birthday present. Before we went to bed, I'd outline for him the problems we had still to solve, and he'd promise to think about them in his half-hour bath in the morning — we always required a bathtub for him in whatever condominium was furnished for us. So it was when we were together.

Fig. 3 In our final years in California, Hans and I were joined by our wives. Shown here, from left to right are Hans, Rose Bethe, Betty Brown, and me. The photo was taken in 1999, at the 60th birthday celebration of nuclear theorist and colleague Chun Wa Wong.

During a massive, hour-and-a-half breakfast of various sliced leftover meats, red currant jam, numerous hot rolls, and lots of tea, we would discuss how to attack the day's problems. Then Hans would estimate how far we could get, and we would go to our office. He usually identified a piece of the problem that he could do before lunch, then filled the pages of blank paper on the top left-hand part of his desk at a steady pace. He was really cross if he needed to work until 10 minutes past noon to finish his preassigned part of the problem, since he clearly desired lunch. But I cannot remember that we ever got severely stuck for long. In the evening we compared solutions; his had been calculated with his slide rule and mine with my $12 calculator. Since I was 20 years and 20 days younger than Hans, I could just about keep up with him!

At the end of January 1996, the day before I was to return to Stony Brook, Kip Thorne came to the office Hans and I shared at Caltech. He said the production of gravitational waves that LIGO (the Laser Interferometer Gravitational-Wave Observatory) should measure from coalescing neutron stars had been estimated by several research workers theoretically, but mergers of neutron stars and black holes had not received as much attention. Hans and I, he continued, were good at calculating things that had not been seen. Could we calculate the contribution of those less-studied mergers?

The previous year I had found that the accepted scenario for evolving a neutron-star binary resulted in a binary comprising a neutron star and a low-mass black hole.[3] So I knew that binaries with a neutron star and a black hole would give at least an order of magnitude more mergers than neutron-star binaries. I said to Hans, "You've now calculated Roman numeral VI of your theory of supernovae. It's time to change." After all, he was only 90. I went on to say that we now had a topic to work on next year at Caltech. He replied, "Oh, no! I want to begin now." So after I got back to Stony Brook, I sent him a 20-page paper by Evert Meurs and Edward van den Heuvel.[4] A few days later, I received from him a page-and-a-half fax, in which he'd derived their main conclusions. "I don't seem to have done it as accurately as they did, but I got the same results," he noted.

We wrote a paper, "Evolution of Binary Compact Objects that Merge," that I think is the best thing Hans and I did.[5] The paper appears in about the middle of the collected works[6] that I edited along with Hans and Chang-Hwan Lee. It and the papers that followed are mostly "about the future," like Ejlert Lövborg's manuscript in Henrik Ibsen's great play *Hedda Gabler*. But since Lee coauthored several of those papers, they cannot be thrown into the fire and burned; in any case, he has computer backups of them.

Hans and I made lots of predictions that only the future will test. The chief among those will not be checked for at least a decade, when LIGO II is completed: We predict that LIGO II will find 20 times more mergers of low-mass black-hole, neutron-star binaries than of neutron-star binaries.

A Large Piece of Life

I hope to see our predictions come true. I am on a low-fat, chiefly fish diet in order to live until LIGO II can test our prediction about mergers of neutron stars and black holes. One of the key ingredients that went into the merger prediction was a result we had obtained for the maximum mass of a neutron star. I must admit I had to do a bit of selling to convince Hans of a strangeness-condensation calculation[7] that we worked out in 1994. Based on that calculation, we concluded that the maximum neutron star mass was about $1.5M_\odot$. But Hans believed the result, and the next year we calculated a comparable maximum mass starting from the approximately $0.75M_\odot$ of nickel production in SN1987a that we were certain went into a black hole.[8]

My close friend Marten van Kerkwijk, who is an excellent observational astronomer, keeps asking me if I still believe the neutron-star mass limit. When I firmly say, "Yes," he responds, "Good," because he thinks it's lovely for observers to be able to refute theoretical predictions. Marten's measured mass of $1.86M_\odot$ for the neutron star Vela X-1 exceeds our limit even when statistical errors are accounted for,[9] but he would be the first to admit that the binary is messy, and notes that "no firm constraints on the equation of state are possible, since systematic deviations in the radial velocity curve do not allow us to exclude a mass around $1.4M_\odot$ as found for other neutron stars." In any case, Marten has been absolutely crucial to the whole program that Hans, Chang-Hwan, and I had been working on. He has checked most of our papers and helped us to avoid published errors. To have our work critiqued by such a skeptical expert as Marten can only be stimulating.

Hans had a particular warm side that he exhibited to his family and to friends such as Rudi Peierls and Viki Weisskopf, as well as to graduate students and colleagues. I was the young interloper who crassly broke into this honored Circle — and he loved it. Following my talk at the 1993 symposium "Celebrating 60 Years at Cornell with Hans Bethe," Hans said, "I knew we had a lot of fun working on the supernova problem, but I didn't know that we had that much fun!" But revealing the warmer part of his character must be assigned to the angels because he would hate me becoming maudlin. As evidence, I note that after I had written a several-page dedication to his wife for the collected works I mentioned earlier, Hans shortened it to "We dedicate this collection of papers to Rose Bethe, who throughout their long evolution fed us, walked us, consoled us and cheered us on."

The two of us worked together right up to his death. In fact, on the phone the morning of the day he died, I told him that C. N. Yang would prepare an article for the centennial volume of *Physics Reports* that I was putting together for Hans's 100th birthday, 2 July 2006.

On the night in 1995 when Rudi Peierls died, Hans had phoned me and said, "A large piece of my life — and yours — has gone." The night that Hans died another large part — father figure, collaborator, and friend — was taken from my life, which is greatly diminished without him.

This article is based in part on my commentaries to Chapters 1 and 12 of Reference 6. I thank World Scientific for permission to use them.

References

1. T. J. Mazurek, J. M. Lattimer and G. E. Brown, *Astrophys. J.* **229**, 713 (1978).
2. H. A. Bethe and P. Pizzochero, *Astrophys. J.* **350**, L33 (1990).
3. G. E. Brown, *Astrophys. J.* **440**, 270 (1995).
4. E. Meurs and E. P. J. van den Heuvel, *Astron. Astrophys.* **226**, 88 (1989).
5. H. A. Bethe and G. E. Brown, *Astrophys. J.* **506**, 780 (1998).
6. H. A. Bethe, G. E. Brown and C.-H. Lee, eds., *Formation and Evolution of Black Holes in the Galaxy: Selected Papers with Commentary*, World Scientific, 2003.
7. G. E. Brown and H. A. Bethe, *Astrophys. J.* **423**, 659 (1994).
8. H. A. Bethe and G. E. Brown, *Astrophys. J.* **445**, L129 (1995).
9. O. Barziv, L. Kaper, M. H. van Kerkwijk, J. H. Telting and J. van Paradijs, *Astron. Astrophys.* **377**, 925 (2001).

Bethe's Hypothesis

Chen Ning Yang and Mo-Lin Ge

Early History

In 1931 Bethe solved,[1] rigorously, the spin wave collision problem of Bloch[2] in one dimension. He used an Ansatz about the algebraic structure of the wave function for the spin waves. This Ansatz has now blossomed into several active fields of research in physics and in mathematics

Before the Second World War the most important development along this line was in the work of Hulthen[3] in 1938. After the war, many papers used Bethe's hypothesis[4]: Orbach (1958), Walker (1959), Lieb, Schultz and Mattis (1961), des Cloizeaux and Pearson (1962), Katsura (1962), Lieb (1963), Lieb and Liniger (1963), Griffiths (1964), McQuire (1964) and others.

In 1966 C. N. Yang and C. P. Yang published three papers[5] on what is now called the XXZ model. They used Bethe's Ansatz and found it very powerful. So they thought it should be explicitly named to honor Bethe. Thus their first paper had the title:

> One-Dimensional Chain of Anisotropic Spin-Spin Interactions
> I. Proof of Bethe's Hypothesis for Ground State in a Finite System

That was where the name Bethe's Hypothesis had first appeared in print. Happily it has now taken hold as Bethe's Hypothesis, or Bethe's Ansatz, or simply BA.

We do not know how Bethe originally had arrived at his hypothesis. It is, in spirit, a generalization of the "only reflection" idea in the classical method of images, which is very old. In the mid-nineteenth century, "Electric Images" was a fashionable subject. In particular William Thomson was proud of his discovery[6] of images in spherical conductors. Maxwell in his *Treatise* wrote

> An electrical image is an electrified point or system of points on one side of a surface which would produce on the other side of that surface the same electrical action which the actual electrification of that surface really does produce.

In Bethe's hands, "only reflection" became generalized to "only reflection and transmission," i.e. "no diffraction." By including transmission the idea becomes more subtle, more complex and much more general, and that is the magic of Bethe's Hypothesis.

Could there be a case in 2 dimensions where the "no diffraction" hypothesis could apply? We have searched for such a case, but without success.

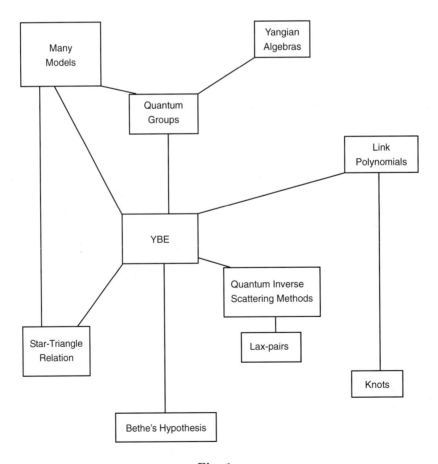

Fig. 1

Later Developments

In retrospect, the key idea in Bethe's Hypothesis was in fact also latent in the star-triangle relations of statistical mechanics, and in Artin's knot theory. These developments led to the Yang-Baxter Equation (YBE). Many solvable models were invented using ideas based on Bethe's Hypothesis and its generalizations throughout the 1970's. Then in the 1980's the idea of "quantum groups" led to explosive advances, especially by the Leningrad School, by the Japanese scientists, and by Drinfel'd. Around the same time Jones and others, starting from the algebraic side, made great advances in knot theory.

In Fig. 1 we present a schematic diagram of these developments in approximate time ordering. It is impossible to list all the key references in these intensive and extensive investigations in mathematics and in physics, so we shall not try to.

References

1. H. A. Bethe, *Z. Physik* **71**, 205 (1931).
2. F. Bloch, *Z. Physik* **61**, 206 (1930); **74**, 295 (1932).
3. L. Hulthen, *Arkiv Mat. Astron. Fysik* **26A**, No. 1 (1938).
4. R. Orbach, *Phys. Rev.* **112**, 309 (1958); L. R. Walker, *Phys. Rev.* **116**, 1089 (1959); E. Lieb, T. Schultz and D. Mattis, *Ann. Phys. (N.Y.)* **16**, 407 (1961); J. des Cloizeaux and J. J. Pearson, *Phys. Rev.* **128**, 2131 (1962); S. Katsura, *Phys. Rev.* **127**, 1508 (1962); E. Lieb, *Phys. Rev.* **130**, 1616 (1963); E. Lieb and W. Liniger, *Phys. Rev.* **130**, 1605 (1963); R. B. Griffiths, *Phys. Rev.* **133**, A768 (1964); J. B. McQuire, *J. Math. Phys.* **5**, 622 (1964).
5. C. N. Yang and C. P. Yang, *Phys. Rev.* **150**, 321 (1966); *ibid.* **150**, 327 (1966); *ibid.* **151**, 258 (1966).
6. W. Thomson, *Cambridge and Dublin Math. J.* **5**, 1 (1850).

Hans Bethe's Contributions to Solid-State Physics

N. David Mermin and Neil W. Ashcroft

Hans Bethe's doctoral research was primarily in solid-state physics. During the late 1920's and early 1930's he played a major role in developing the new quantum theory of solids. Though nuclear physics became his main interest in the mid 1930's, he continued to write papers in solid-state physics into the late 1940's, and remained interested in the subject all his life.

Hans Bethe started his scientific career in 1927 as a solid-state physicist. Although he said[1] that by 1933 "my real interest was in nuclear physics," he continued to make major contributions to the theory of solids over a twenty-year span, publishing his last paper on the subject in 1947. In 1933 he was regarded as a solid-state physicist by no less a figure than Wolfgang Pauli, as revealed in an anecdote that Victor Weisskopf loved to tell.[2]

When Weisskopf arrived in Zurich in the fall of 1933 to assume his duties as Pauli's assistant, Pauli's first words to the new postdoc (after "Who are you?") were "I wanted to take Bethe, but he works on solid-state theory, which I don't like, although I started it." Pauli did indeed make the first application of Fermi-Dirac statistics to electrons in metals, explaining their anomalously low contribution to the observed paramagnetic susceptibility. He thereby inspired Bethe's thesis advisor, Arnold Sommerfeld, to extend this approach to the computation of other electronic ground-state and thermal properties of metals. Sommerfeld then set the young Hans Bethe to work on other aspects of the burgeoning new quantum theory of solids. The punch line of Weisskopf's anecdote reveals the reputation Bethe had already acquired at the age of 27. Weisskopf reports that "Pauli then gave me some problem to study ... and after a week he came and asked me what I did. I showed him and he said, 'I should have taken Bethe.'"

Bethe was not quite twenty years old when he went to Munich to be a graduate student with Sommerfeld. Schrödinger's papers were just coming out and, according to Bethe[1] "that was the sort of thing, differential equations with eigenvalues, that Sommerfeld had loved for a decade or two. And he gave a special course for graduate students on the differential equations of physics, and Schrödinger fitted right in. So we all were asked to take one section of Schrödinger and give a seminar on it. So that's how I learned some of Schrödinger's perturbation theory, because that was the section I was to give in the seminar."

In an essay, "Reminiscences on the Early Days of Electron Diffraction,"[3] Bethe adds that "Ever since that time I have been very fond of perturbation theory in all its forms." He then goes on to tell what happened next. "The Davisson-Germer paper reporting the observation of electron diffraction appeared, confirming the ideas of de Broglie and Schrödinger and mak[ing] the entire subject much more real." Sommerfeld suggested to Bethe that he look into why the diffraction maxima did not occur at the right incident wavelengths, and Bethe quickly discovered that this could easily be remedied by taking into account the change in potential energy suffered by the electrons as they entered the crystal. He reported this in his very first publication[4] and was subsequently "rather dismayed" to learn that he had been scooped by the X-ray crystallographer A. L. Patterson — "a great disappointment for a physicist who has just published his first paper."

Interestingly, Patterson has quite a different memory of this "disappointment." He reports[5] that immediately after receiving and studying the Davisson-Germer paper he came up with "an incorrect interpretation of why the diffraction angle did not check with the lattice spacing. I interpreted this as a change in lattice spacing near the surface of the crystal whereas Bethe correctly explained it a few weeks later in terms of a refractive index effect."

Scooped or not scooped, life goes on. Sommerfeld, following Pauli, had found that electrons in metals have quite a large kinetic energy in their ground state. Bethe suggested to Sommerfeld that the work function of a metal could be understood as the combination of his own negative potential energy and Sommerfeld's positive kinetic energy. "After this suggestion, Sommerfeld considered me a great expert in solid-state physics, which I was not. But his favorable estimate led to our joint article on the electron theory of metals[3] in the *Handbuch der Physik*, Vol. XXIV, a report which gave me a lot of pleasure."

Before they turned to the Handbuch article, Sommerfeld suggested to Bethe that he produce a more detailed theory of electron diffraction by crystals, following along the lines developed by P. P. Ewald's 1917 work on the dynamical theory of X-ray diffraction. Bethe soon discovered that this was quite an easy job, since Ewald had to deal with a vector field, while the field associated with "de Broglie waves" was scalar. "Having studied all these complications [associated with the vector field] in Ewald's paper, I was only too happy to discard them, and to retain only his fundamental idea."[3] Ewald's idea was to expand the spherical wave scattered by each atom in plane waves.

Bethe was thus the first to work out, in the context of electron scattering by a crystal, what is now known as the nearly-free-electron approach to electronic band structure.[6] He even found, independent of Felix Bloch and at about the same time, according to Hoddeson *et al.*,[7] that the wave function inside the crystal had the form of a plane wave modulated by a function with the periodicity of the lattice of ions ("Bloch's theorem"). Because his focus was on the Bragg-reflected non-propagating solutions, existing in the energy gaps, rather than on the propagating solutions in the allowed bands, his work did not play a direct role in the development of the band-theoretic explanation of metallic conduction, though it seems to have had an impact on the way he thought about these questions with his good friend and fellow graduate student Rudolf Peierls.[1]

But Bethe seems not to have fully appreciated the broader implications of his thesis for the free propagation of electrons in a perfect crystal until after the work of Bloch, which approached the problem not from the free-electron side, but from the other extreme — what is now called tight-binding. In Ref. 1 Bethe remarks that neither Sommerfeld nor Drude "had proved that the electrons indeed were free. That was only done by Bloch ... [who] proved that the wavefunction of an electron was a plane wave multiplied by a function which was periodic with the period of the lattice, and Bloch, in a way, was the fundamental origin of the free electron picture .... [T]o me Bloch was essential." On the other hand, when asked whether he was suspicious of Sommerfeld's treatment of electrons as free before Bloch's work had appeared, he responded "Not at all. I came from a slightly different direction. I was told by Sommerfeld, for my thesis, to make a theory of the experiments of Davisson and Germer which proved the wave nature of electrons, and proved that electrons behaved very much like X-rays. And so it was obvious to me that electrons were free."

That Bethe was on top of all major developments in the formative years of modern solid-state physics is evident from his comprehensive treatise on the quantum theory of solids — the very first textbook in the field. This is the joint article Sommerfeld had suggested, which appeared as a review, almost three hundred pages long, "Elektronentheorie der Metalle."[8] There is a footnote attached to Sommerfeld's author-identification stating that "Nur für das erste Kapitel dieses Berichtes bin ich verantwortlich. Verdienst und Verantwortung an den folgenden Kapiteln kommen H. Bethe zu." (I am responsible only for the first chapter of this report. Credit and responsibility for the chapters that follow it belong to H. Bethe.)

Sommerfeld's part of the project, a little over 10 per cent of the book (in which form it was reissued by Springer Verlag in 1967), discusses his refinement of the free electron model of Drude, in which the electrons continue to obey classical mechanics but their equilibrium velocity distribution, following Pauli's first such application, is given by quantum Fermi-Dirac statistics, rather than classical Maxwell-Boltzmann statistics. The remaining 90 per cent of this extensive overview of the quantum mechanics of electrons in solids was produced by Bethe in a single year — a year in which his primary interests, stimulated by Chadwick's 1932 discovery of the neutron, had already started to shift to nuclear physics.[1] Although the quantum theory of solids, and indeed, quantum mechanics itself, was less than a decade old, Bethe's review is remarkably deep, broad, and modern in its approach, setting a clear course for the next generation of textbooks to the point where one wonders why the authors of future texts bothered to undertake their projects. "Shows the stability of the theory" was Hans's laconic comment, when one of us remarked on this in 2004.[1]

In addition to his thesis and his comprehensive review of the subject, Bethe made several other major contributions to solid-state physics. His monumental article on the splitting of atomic levels by crystal fields[9] (an English translation can be found in his *Selected Works*[10]), unlike his thesis which works in the nearly free electron limit, begins in the opposite limit of tight binding. Here the electronic states locally resemble those of a free atom, subject to perturbations whose origin and symmetry reflect its surroundings in the crystal. Levels, which for an isolated atom would possess degeneracies under the full rotation group, split when the atom's environment has the reduced symmetry of the point group appropriate to its crystalline environment. In this very first application of group-representation theory to solid-state physics, Bethe gives the problem his characteristically thorough airing, determining the splittings and the residual degeneracies in the crystalline environment for an enormous number of experimentally accessible cases. He even calculates the character tables from first principles.

Later Bethe characterized this vast piece of work — a treatise on a major application of group-representation theory — as something he did "only because I had studied a book on group theory, and you can't really understand something unless you apply it and work with it yourself. So since Wigner had done all the really important things with group theory, I thought the only thing that remains to be done is to take an atom in a crystal of various symmetries and see how the energy levels will look there. I am told that people have used that paper, but I have never seen what has come out of it."[11] Among the many things that came out of it were a second paper of his own, that extended the method to the presence of magnetic fields, to explain features of the Zeeman effect in the rare-earth salts,[12] and a third paper he wrote with F. H. Spedding, applying the method to the absorption spectra of rare-earth salts.[13]

Superconductivity, discovered in 1911 in the Leiden laboratory of H. Kamerlingh-Onnes, but not explained until 46 years later, by Bardeen, Cooper, and Schrieffer, was already considered a serious unsolved problem for the quantum theory of solids in the late 1920's. Bethe did not attempt a theory of superconductivity, but he was interested enough to publish criticisms of unsuccessful attempts by others. Among these is his comment[14] on work by R. Schachenmeier — an article which, except for a short opening paragraph, a concluding sentence, and a brief final summary, is written entirely in small print. The explanation for this unusual choice of font is that Schachenmeier had published a response to criticism of his work in the Sommerfeld-Bethe treatise, complaining, among other things, that the authors (actually Bethe) had relegated his work to a section in small print. By replying in a tiny font, Bethe solved the problem of how to put on record his reply to Schachenmeier, without drawing even more attention to work he clearly regarded as without merit. The concluding sentence (in normal print) notes that "Die hier gegebene Aufzählung der Irrtümer in den Arbeiten von Schachenmeier macht keinen Anspruch auf Vollständigkeit." (The enumeration given here of the errors in the work of Schachenmeier makes no claim to be complete.)

In 1931, during his year in Rome with Fermi, Bethe became interested in extending Heitler and London's theory of the hydrogen molecule to a whole chain of atoms — a one-dimensional model of a metal. In contemporary terms he examined the one-dimensional Heisenberg model with ferromagnetic nearest-neighbor coupling, and found a complete solution for all the eigenvalues and eigenstates, thereby improving on the spin-wave approximation of Bloch. This is the work[15] in which the celebrated "Bethe ansatz" is introduced, as expanded on elsewhere in this volume by C. N. Yang. The peculiar title of the paper ("Zur Theorie der Metalle. I") finds its explanation in the next to last paragraph, where Bethe optimistically promises a future paper in which "this method will be extended to space lattices, and its physical implications for cohesion, ferromagnetism and electrical conductivity will be derived." To this day no one has managed to accomplish the extension to higher dimensions.

The year 1931 also saw the publication of a short note,[16] which Bethe was sufficiently fond of to choose as one of the 28 entries (out of 290 candidates) in his *Selected Works*.[10] "On the Quantum Theory of the Temperature of Absolute Zero," written with G. Beck and W. Riezler, purports to calculate $-273°$ Celsius as $-(2/\alpha - 1)$ degrees, where α is Sommerfeld's fine structure constant. This spoof of Eddington aroused the ire of the editor of *Die Naturwissenschaften*, who apparently took it at face value and demanded that the authors provide an erratum. This states that the article "was not meant to be taken seriously. It was intended to characterize a certain class of papers in theoretical physics of recent years which are purely speculative and based on spurious numerical agreements. In a letter received by the editors from these gentlemen they express regret that the formulation they gave to the idea was suited to produce misunderstanding." If Bethe felt it belonged among his selected papers, then it certainly belongs in our survey of his contributions to solid-state physics, for it begins: "Let us consider a hexagonal lattice."

His activity in solid-state physics did not cease after Bethe's interest had shifted to nuclear physics. In 1935, he reports,[11] a "problem was suggested to me by Sir Lawrence Bragg, while I was in Manchester," in response to which he "put just a very slight amount of respectability into the [1934] theory of Bragg and Williams on the way order and disorder come about" in alloys. Bragg and Williams had studied the transition between the ordered and disordered phases of a binary alloy using a very simple mean field approximation, in which the interaction energy of an atom at any site of one sublattice with the specific configuration of its immediate neighbors on the other sublattice is replaced by the mean value of that energy over all such nearest-neighboring pairs. The Bragg-Williams approximation is independent of any features of the lattice geometry and gives ordering even in one dimension, where one knows from quite simple arguments that it cannot occur.

Bethe's refinement[17] treats exactly the interaction of an atom at any given site with the actual configuration of atoms at its nearest-neighboring sites, introducing a mean field approximation only to treat the interaction of those nearest neighbors with their own additional neighbors. This allows lattice geometry to enter the final results. The transition temperature now depends on the number of nearest neighbors and a transition no longer occurs in one dimension. This improvement on Bragg and Williams is sometimes called the Bethe-Peierls approximation, which may reflect the fact that Bethe gives "special thanks" to "my friend Dr. Peierls, who gave innumerable valuable suggestions. In fact [the central calculation of the paper] is entirely due to him." Peierls also published work of his own on the problem, at about the same time.

After arriving at Cornell in 1935 (at the age of 29), Bethe continued to write about phase transitions in metals, although by then his activity in nuclear physics and, by the late 1930's, nuclear astrophysics, was intense. Robert Marshak[18] describes "a conference on solid-state physics which Bethe had organized at Cornell" at about the time Marshak went there to become his graduate student in 1937. We say more about this little known "Symposium on the Structure of Metallic Phases" in an appendix. Bethe contributed a paper to the Symposium, a beautifully pedagogical introduction to the theory of cooperative phenomena and its application to order-disorder transitions in alloys, a much longer version of which he published in 1938.[19] The great Cornell chemist J. G. Kirkwood talked at the Symposium about closely related work of his own, and in 1939 they published a joint article, comparing their two approaches.[20]

In 1960 Cyril Domb pointed out that the Bethe approximation is exact on lattices that have no closed circuits of nearest-neighbor bonds (Cayley trees). He named them "Bethe lattices,"[21] thereby giving birth to a terminology now widely used in condensed matter physics, where a great variety of problems can be solved exactly on Bethe lattices.

Bethe's 1951 paper with Edwin Salpeter[22] is a contribution to relativistic quantum field theory, which they apply to make a point about the relativistic corrections to the binding energy of the deuteron. It must nevertheless be mentioned in any survey of Bethe's contributions to solid-state physics, because the integral equation Salpeter and Bethe derive, expressing the two-particle propagator as a sum of ladder diagrams, has been known ever since to generations of enthusiastic solid-state theorists as the "Bethe-Salpeter equation" (in blatant disregard for the order of names on the paper — yet another example of the regrettable Matthew effect).

Though not in solid-state physics and never published in a scientific journal, Bethe's 1941 report with Edward Teller, "Deviations from thermal equilibrium in shock waves,"[23] must also be mentioned. Bethe regarded it highly enough to include among the 10% of his papers appearing in his *Selected Works*. He explains in his commentary on their report that "after the fall of France to the Nazis in June 1940, Teller and I felt that we wanted to contribute to the Allied War Effort." This was "before they had any official connections with Government agencies during World War II," as a preface to a reissued version of the report gently puts it — at that point Bethe was still an "enemy alien." So they asked Theodor von Kármán what they could do on their own to help, and in response to his suggestion produced an important paper (quickly classified as secret) on the physics of shock waves.

Their main result was that the asymptotic values of the local mean velocity, density, pressure, and temperature "on the high pressure side at sufficient distance from the front of the shock wave are uniquely determined by the values of these quantities on the low pressure side," independent of microscopic details underlying the grossly nonequilibrium conditions prevailing in the shock front itself. Though derived for shock waves in the atmosphere, this general result turned out to follow from little more than the conservation laws, remaining valid in solids and liquids as well as in dilute gases. It remains one of the key papers in the study of solids far from equilibrium.

Bethe's final paper that is unambiguously and explicitly in the area of solid-state physics appeared in 1947. It expands on the celebrated 1933 calculation by Wigner and Seitz of the electronic wave-functions for sodium. This evidently impressed Bethe enough for him to write an early report on their approach.[24] Fourteen years and a world war later he returned to the problem with his student Fred Von der Lage.[25] Wigner and Seitz had noted that a major difference between the atomic and metallic problems was that the boundary condition of vanishing wave-function at infinity appropriate to the atom, is replaced in the metal by a boundary condition of vanishing normal derivative on the surface of the primitive cell with the full symmetry of the crystal — what is now called the Wigner-Seitz cell. To make things more tractable they replaced this cell by a sphere of the same volume.

How to improve on that must have been obvious to the man who first applied group theoretic methods to electronic levels in a crystal. The Bethe-Von der Lage paper applies the boundary conditions at points of high symmetry at the boundary of the actual Wigner-Seitz cell. The analytical key is to expand the wavefunctions in "Kubic harmonics" — linear combinations of spherical harmonics transforming under the appropriate irreducible representations of the cubic group. Their calculation showed that the conduction electrons in the first few Brillouin zones of sodium (at ordinary pressures and temperatures) have energies that are very close to the free electron values.

The final paragraph of Ref. 25 begins with the plaintive remark that "No experimental verification of sufficient resolving power exists to support these results," and indeed, it took another 20 years before their general conclusions were confirmed by detailed precision measurements of the Fermi surface of sodium. This appears to be a manifestation of a broader frustration Bethe felt with solid-state physics in the early days, which may have played as large a role as the neutron in accounting for his departure from the field. In his Trieste lecture[11] he remarks that "I worked on solid state but I must say at the time it was a far less satisfactory pursuit than nuclear physics. It was really much too early to do solid state seriously. My ambition at the time was to calculate such things as the shape of the Fermi surface for the electrons in silver or at least in sodium, and then to have some experimental confirmation for this. By now we know the Fermi surfaces of these substances, but at that time there was absolutely no way of doing it. All you could measure was conductivity and a few thermoelectric and magnetic effects, gross numbers which certainly would never give you any information of the kind I wanted."

This seems to us characteristic of the man. Brilliant successes with model problems — and in his brief career as a solid-state physicist Bethe had more of these than most of us achieve in a lifetime — were for him just a diversion from the primary task of physics, which was to calculate (preferably with a slide rule) real quantitatively measurable properties of real physical systems: atoms, solids, nuclei, stars. His early achievements in solid-state physics were more conceptual than quantitative, and for him, although this was a source of great pleasure, it may also have been tinged with disappointment.

Nevertheless he remained interested in solid-state and then condensed-matter physics all his life. NDM recalls that in the mid 1970's, shortly after the discovery at Cornell of superfluidity in helium-3, he encountered Bethe in an elevator. "f-wave pairing!," Hans said like an oracle, "it's probably f-wave pairing." The conventional wisdom at that point had coalesced around p-wave pairing, but for no compelling reason other than that it was easier to deal with. Inspired by this encounter, NDM started exploring some of the simpler consequence of f-wave pairing, and found one that was flatly contradicted by a recent nuclear-magnetic-resonance experiment. As far as he knows, it remains to this day the most direct experimental evidence against f-wave pairing. He would never have stumbled on it if he hadn't found himself in that elevator with his senior colleague, Hans Bethe.

Appendix: The 1937 Cornell Symposium

We were intrigued by the remark of Robert Marshak[18] that Hans Bethe had organized a conference at Cornell on solid-state physics in or around 1937. We were unaware of such an event, and would have thought that by then, in spite of his occasional publications in the field, Bethe was far too preoccupied with nuclear physics for such an undertaking. Our own efforts to learn about it led nowhere. But Patricia Viele of the Cornell Physical Sciences Library discovered in the University Archives a printed announcement of a 1937 *Symposium on the Structure of Metallic Phases* that gives the program of talks, together with a set of brittle mimeographed pages that appear to be the texts of those talks.

Having learned the name of the conference, we were then able to uncover two footnotes citing the Symposium in the *Physical Review*, in papers by William Shockley (1937) and Francis Bitter (1938). There is also a detailed announcement of the Symposium in the June 4, 1937 issue of *Science*.[26] Otherwise it does not seem to have left much of a mark on the scientific literature or the history of science.

The Symposium took place July 1–3, 1937, in the physics building, Rockefeller Hall. It was no casual event, the address of welcome being given by the President of the University, Edmund Ezra Day. We were impressed by the registration fee ($1) and the cost of accommodation in Balch Hall ($2 a night or $5 for all three nights). But most of all we were impressed by the mimeographed set of Symposium papers,[27] whose titles and authors are as follows:

Thermodynamical Considerations of Order and Disorder — J. C. Slater, Massachusetts Institute of Technology

Statistical Theory of Cooperative Phenomena — H. A. Bethe, Cornell University

Order and Disorder in Solids — J. G. Kirkwood, Cornell University

Superstructures in Alloy Systems — F. C. Nix, Bell Telephone Laboratories

Phase Changes of the First Kind — E. R. Jette, Columbia University

Kinetics of Reactions in Solid Alloys — R. F. Mehl, Carnegie Institute of Technology

Quantum Theory of Metals — F. Seitz, General Electric Company

The Ferromagnetic Properties of Alloys — F. Bitter, Massachusetts Institute of Technology

Magnetic Interaction in Crystals — L. W. McKeehan, Yale University

Directional Ferromagnetic Properties of Cubic Metals — R. M. Bozorth, Bell Telephone Laboratories

These differ from the titles and speakers listed in the program in two ways. The program lists at the end "Critical Comments and Summarizing Remarks" by J. C. Slater. And neither the program nor the announcement in *Science* mention a talk by Bethe. The program does list a talk with the title of Bethe's paper, but it is under the name of J. C. Kirkwood and described as a "report prepared in collaboration with H. A. Bethe, Department of Physics, Cornell University." Since the program lists no talk with the title of Kirkwood's paper, it seems possible that Kirkwood delivered both Bethe's paper and his own in a single talk. On the other hand in the manuscript of his opening talk, Slater looks ahead to "the paper of Professor Bethe," so we shall have to wait for the historians to settle the question of what actually took place.

What is not a matter of doubt is that the manuscripts for the Symposium provide a fascinating and highly readable glimpse into the state of the theory of phase transitions in solids, as well as the broader field of solid-state physics, at that still relatively early stage of its development.

References

1. H. A. Bethe and N. David Mermin, "A Conversation about Solid-State Physics," *Physics Today*, June 2004, pp. 53–56.
2. V. F. Weisskopf, *The Privilege of Being a Physicist*, W. H. Freeman, New York, 1989, pp. 158–159.
3. *50 Years of Electron Diffraction*, eds. P. Goodman and D. Reidel, Dordrecht, Boston, London, 1981.

4. H. A. Bethe, *Naturwissenschaften* **15**, 787 (1927).
5. A. L. Patterson, in *50 Years of X-Ray Diffraction*, ed. P. Ewald, *International Union of Crystallography*, 612 (1962).
6. H. A. Bethe, *Ann. d. Physik* **87**, 55–129 (1928).
7. L. Hoddeson, G. Baym and M. Eckert, *Rev. Mod. Phys.* **59**, 287–327 (1987).
8. A. Sommerfeld and H. A. Bethe, *Handbuch der Physik* **24**, Part II (1933).
9. H. A. Bethe, *Ann. d. Phys.* **3**, 133–208 (1929). English translation in Ref. 10.
10. *Selected Works of H. A. Bethe: With Commentary*, World Scientific, 1997.
11. H. A. Bethe, in "From a Life of Physics: Evening Lectures at the International Centre for Theoretical Physics, Trieste, Italy," special supplement of the International Atomic Energy Agency (IAEA) Bulletin, Vienna, 1969.
12. H. A. Bethe, *Zeitschrift für Physik* **60**, 218–233 (1930).
13. H. A. Bethe and F. H. Spedding, *Phys. Rev.* **52**, 454–455 (1937).
14. H. A. Bethe, *Zeitschrift für Physik* **90**, 674–679 (1934).
15. H. A. Bethe, *Zeitschrift für Physik* **71**, 205–226 (1931). English translation in Ref. 10, pp. 155–183.
16. G. Beck, H. A. Bethe and W. Riezler, *Die Naturwissenschaften* **19**, 39 (1931). English translation in Ref. 10.
17. H. A. Bethe, *Proc. Roy. Soc. A* **150**, 552–575 (1935).
18. R. E. Marshak, remarks introducing Hans A. Bethe in Ref. 11.
19. H. A. Bethe, *J. Appl. Phys.* **9**, 244–251 (1938).
20. H. A. Bethe and J. G. Kirkwood, *J. Chem. Phys.* **7**, 578–582 (1939).
21. C. Domb, *Adv. Phy.* **9**, 149–361 (1960).
22. E. E. Salpeter and H. A. Bethe, *Phys. Rev.* **84**, 1232–1242 (1951).
23. H. A. Bethe and E. Teller, unpublished report, reprinted in Ref. 10, pp. 297–345.
24. H. A. Bethe, *Helvetica Physica Acta* **VII**, Sup. II, 18–23 (1934).
25. F. C. Von der Lage and H. A. Bethe, *Phys. Rev.* **71**, 612–622 (1947).
26. *Science* **85**, No. 2214, June 4, 1937, pp. 538–539.
27. Readers interested in learning more about the Symposium manuscripts should send an email to the Division of Rare and Manuscript Collections, Cornell University Library, *rareref@cornell.edu*, referring to the Symposium on the Structure of Metallic Phases, held at Cornell, July 1–3, 1937.

Hans Bethe and the Nuclear Many-Body Problem

Jeremy Holt and Gerald E. Brown

1. The Atomic and Nuclear Shell Models

Before the Second World War, the inner workings of the nucleus were a mystery. Fermi and collaborators in Rome had bombarded nuclei by neutrons, and the result was a large number of resonance states, evidenced by sharp peaks in the cross section; i.e., in the off-coming neutrons. These peaks were the size of electron volts (eV) in width. (The characteristic energy of a single molecule flying around in the air at room temperature is the order of 1/40 of an electron volt. Thus, one electron volt is the energy of a small assemblage of these particles.) On the other hand, the difference in energy between low-lying states in light and medium nuclei is the order of MeV, one million electron volts.

Niels Bohr's point[1] was that if the widths of the nuclear levels (compound states) were more than a million times smaller than the typical excitation energies of nuclei, then this meant that the neutron did not just fall into the nucleus and come out again, but that it collided with the many particles in the nucleus, sharing its energy. According to the energy-time uncertainty relation, the width ΔE of such a resonance is related to the lifetime of the state by $\tau = \hbar/\Delta E \sim \hbar/(1\,\text{eV}) \simeq 10^{-15}$ sec. In fact, the time $t = \hbar/(10\,\text{MeV}) \simeq 10^{-22}$ sec is the characteristic time for a nucleon to circle around once in the nucleus, the dimension of which is Fermis (1 Fermi = 10^{-13} cm). Thus, a thick "porridge" of all of the nucleons in the nucleus was formed, and only after a relatively long time would this mixture come back to the state in which one single nucleon again possessed all of the extra energy, enough for it to escape. The compound states in which the incoming energy was shared by all other particles in the porridge looked dauntingly complicated.

At the time, atomic physics was considered to be very different from nuclear physics, the atomic many-body problem being that of explaining the makeup of atoms. Niels Bohr[2] had shown, starting with the hydrogen atom, that each negatively charged electron ran around the much smaller nucleus in one of many "stationary states." Once in such a state, it was somehow "protected" from spiraling into the positively charged nucleus. (This "protection" was understood only later with the discovery of wave mechanics by Schrödinger and Heisenberg.) The allowed states of the hydrogen atom were obtained in the following way. The Coulomb attraction between the electron and proton provides the centripetal force, yielding

$$\frac{e^2}{r^2} = \frac{mv^2}{r},\qquad(1)$$

where $-e$ is the electron charge, $+e$ is the proton charge, m is the electron mass, v is its velocity, and r is the radius of the orbit. Niels Bohr carried out the quantization, the meaning of which will be clear later, using what is called the classical action, but a more transparent (and equivalent) way is to use the particle-wave duality picture of de Broglie (Prince L. V. de Broglie received the Nobel prize in 1929 for his discovery of the wave nature of electrons) in which a particle with mass m and velocity v is assigned a wavelength

$$\lambda = \frac{h}{mv},\qquad(2)$$

where h is Planck's constant. If the wave is to be stationary, it must fit an integral number of times around the circumference of the orbit, leading to

$$n\frac{h}{mv} = 2\pi r.\qquad(3)$$

Eliminating v from Eqs. (1) and (3), we find the radius of the orbit to be

$$r = \frac{n^2\hbar^2}{me^2},\quad \text{where}\quad \hbar = \frac{h}{2\pi}.\qquad(4)$$

Using Eq. (1), the kinetic energy of the electron is

$$T = \frac{1}{2}mv^2 = \frac{e^2}{2r}.\qquad(5)$$

The potential energy is

$$V = -\frac{e^2}{r} = -2T, \qquad (6)$$

which follows easily from Eq. (5). Equation (6) is known as a "virial relation," an equation that relates the kinetic and potential energies to one another. In the case of a potential depending on r as $1/r$, the kinetic energy is always equal to $-1/2$ of the potential energy. From this relation we find that the total energy is given by

$$E = T + V = -T. \qquad (7)$$

Finally, one finds by combining Eqs. (7), (5), and (4) that

$$E_n = -\frac{1}{2}\frac{me^4}{n^2\hbar^2} = -\frac{1}{n^2}Ry, \qquad (8)$$

where Ry is the Rydberg unit for energy

$$Ry \simeq 13.6 \text{ electron volts}. \qquad (9)$$

Thus, we find only a discrete set of allowed energies for an electron bound in a hydrogen atom, and we label the corresponding states by their values of n. Since $n = 1$ for the innermost bound orbit in hydrogen, called an s-state for reasons we discuss later, 13.6 electron volts is the ionization energy, the energy necessary to remove the electron.

Electrons are fermions, such that only one particle can occupy a given quantum state at a time. This is called the "exclusion principle." (Wolfgang Pauli received the Nobel prize in 1945 for the discovery of the exclusion principle, also called the "Pauli principle.") Once an electron has been put into a state, it excludes other electrons from occupying it. However, electrons have an additional property called spin, which has the value $1/2$ (in units of $\hbar$), and this spin can be quantized along an arbitrary axis to be either up or down. Thus two electrons can occupy the 1s state. Putting two electrons around a nucleus consisting of two neutrons and two protons makes the helium atom. Helium is the lightest element of the noble gases. (They are called "noble" because they interact very little with other chemical elements.) Since the 1s shell is filled with two electrons, the helium atom is compact and does not have an empty 1s orbital which would like to grab a passing electron, unlike hydrogen which is very chemically active due to a vacancy in its 1s shell.

To go further in the periodic table we have to put electrons into the $n = 2$ orbit. A new addition is that an electron in the $n = 2$ state can have an orbital angular momentum of $l = 0$ or 1, corresponding to the states $2s$ and $2p$, respectively. (The spectroscopic ordering is $s\,p\,d\,f\,g\,h\,i$ for $l = 0 - 6$, a notation that followed from the classification of atomic spectra well before the Bohr atom was formulated.) The angular momentum $l = 1$ of the $2p$ state can be projected on an arbitrary axis to give components $m = 1, 0$, or -1. So altogether, including spin, six particles can be put into the $2p$ state and two into the $2s$ state. Thus, adding eight electrons in the $2s$- and $2p$-states, the next member of the noble gases, neon, is obtained. It is particularly compact and the electrons are well bound, because both the $n = 1$ and $n = 2$ shells are filled.

Consider an element in which the $n = 2$ shell is not filled, oxygen, which has eight electrons, two in the $n = 1$ shell and six in the $n = 2$ shell. Oxygen in the bloodstream or in the cellular mitochondria is always on the alert to fill in the two empty orbits. We call such a "grabbing" behavior "oxidation," even though the grabbing of electrons from other chemical elements is done not only by oxygen. The molecules that damage living cells by stealing their electrons are called "free radicals," and it is believed that left alone, they are a major cause of cancer and other illnesses. This is the origin of the term "oxidative stress." We pay immense amounts of money for vitamins and other "antioxidants" in order to combat free radicals by filling in the empty states.

We need to add one further piece to the picture of the atomic shell model, the so-called j–j coupling, which really gave the key success of the nuclear shell model, as we explain later. The j we talk about is the total angular momentum, composed of adding the orbital angular momentum l and the spin angular momentum s. The latter can take on projections of $+1/2$ or $-1/2$ along an arbitrary axis, so j can be either $l + 1/2$ or $l - 1/2$. The possible projections of j are $m = j, j - 1, \ldots, -j$. Thus, if we reconsider the p-shell in an atom, which has $l = 1$, the projections are reclassified to be those of $j = 3/2$ and $j = 1/2$. The former has projections $3/2, 1/2, -1/2$, and $-3/2$, the projections differing by integers, whereas the latter has $+1/2$ and $-1/2$. Altogether there are six states, the same number that we found earlier, through projections of $m_l = -1, 0, 1$ and of $m_s = +1/2, -1/2$. The classification in terms of j is important because there is a spin-orbit coupling; i.e., an interaction between spin and orbital motion which depends on the relative angle between the spin and orbital angular momentum. This interaction has the effect of increasing the energy of a state for which the spin is in the same direction as the orbital angular momentum, as in the $2p_{3/2}$ state being higher in energy than the $2p_{1/2}$ state. In this notation, the subscript refers to the total angular momentum. The filling of the shells in the j–j classification scheme is the same as in the l scheme in that the same number of electrons fill a shell in either scheme, the only difference being the subshell labels.

The above is the atomic shell model, so called because electrons are filled in shells. The electric interaction is weak compared with the nuclear one, down by a factor of $\alpha = e^2/\hbar c = 1/137$ from the nuclear interaction (called the strong interaction). The atomic shell model is also determined straightforwardly because the nucleus is very small (roughly 10,000 times smaller in radius than the atom), and it chiefly acts as a center about which the electrons revolve. In atoms the number of negatively charged electrons is equal to the number of positively charged protons, the neutrons being of neutral charge. In light nuclei there are equal numbers of neutrons and protons, but the number of neutrons relative to the protons grows as the mass number A increases, because it is relatively costly in energy to concentrate the repulsion of the protons in the nucleus. On the whole, electric forces play only a minor role in the forces between nucleons inside the nucleus, except for determining the ratio of protons to neutrons. By the time we get to ^{208}Pb with 208 nucleons, 126 neutrons and 82 protons, we come to a critical situation for the nuclear shell model, which requires the spin-orbit force, as we discuss later.

After a decade or two of nuclear "porridge," imagine people's surprise when the nuclear shell model was introduced in the late 1940's[3,4] and it worked; i.e., it explained a lot of known nuclear characteristics, especially the "magic numbers." That is, as shells were filled in a prescribed order, those nuclei with complete shells turned out to be substantially more bound than those in which the shells were not filled. In particular, a particle added to a closed shell had an abnormally low binding energy.

What determines the center in the nuclear shell model? In the case of the atomic shell model, the much heavier and much smaller nucleus gave a center about which electrons could be put into shells. Nuclei were known to be tightly bound. There must be a center, and most simply the center would be exactly in the middle of the charge distribution. This charge distribution can be determined by considering two nuclei that differ from one another by the exchange of a proton and a neutron. In this case the two nuclei will have different binding energies due to the extra Coulomb energy associated with the additional proton. This difference could be measured by the energy of the radioactive decay of the one nucleus into the other and estimated roughly as Ze^2/R (where R is the radius of the nucleus and Z is the number of protons). This gave $R \sim 1.5 A^{1/3} \times 10^{-13}$ cm, where A is the mass number. (Later electron scattering experiments, acting like a very high resolution electron scattering microscope, gave the detailed shape of the charge distribution and basically replaced the 1.5 by 1.2.)

The most common force is zero force, that is, matter staying at rest in equilibrium. The force, according to Newton's law, is the (negative) derivative of the potential. If the potential at short distances is some constant times the square of r; i.e., $V = Cr^2$, then the force $F = -dV/dr = -2Cr$ is zero at $r = 0$. Furthermore, the negative sign indicates that any movement away from $r = 0$ is met by an attractive force directed back toward the center, leading to stable equilibrium. Such a potential is called a harmonic oscillator. It occurs quite commonly in nature, since most matter is more or less in equilibrium.

The energy of a particle in such a potential can be expressed classically as

$$E = \frac{p^2}{2M} + \frac{1}{2}M\omega^2 r^2, \qquad (10)$$

where p is the particle's momentum, M is its mass, and ω describes the strength of the restoring force. Increasing ω will bind the nucleons closer together, which allows ω to be related to the nuclear radius R by

$$\hbar\omega \sim \frac{\hbar^2}{MR^2}. \qquad (11)$$

Using harmonic oscillator wave functions with ω determined so that

$$R = 1.2 A^{1/3} \times 10^{-13} \text{cm} \qquad (12)$$

gives a remarkably good fit to the shape of the charge distribution (in which only the protons are included) of the nucleus. The possible energy levels in such a potential are shown in Fig. 1 where we have labeled them in the same way as in the atomic shell model. Note that the distances between neighboring levels are always the same, $\hbar\omega$.[a] We redraw the potential in Fig. 2, incorporating the j–j coupling and a spin-orbit interaction.

---

[a]The situation is analogous to that with light quanta, which have energy $\hbar\omega$ each, so that an integral number of quanta is emitted for any given frequency $\hbar\omega$.

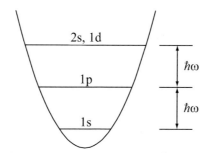

Fig. 1 Harmonic oscillator potential with possible single-particle states.

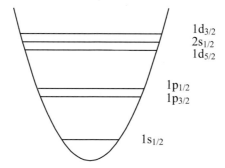

Fig. 2 Harmonic oscillator potential with possible single particle states in j–j coupling.

Note that the spin-orbit splitting, that is, the splitting between two states with the same l, is somewhat smaller than the distance between shells,[b] so that the classification of levels according to l as shown in Fig. 1 gives a good zero-order description.

At first sight the harmonic oscillator potential appears unreasonable because the force drawing the particle back to the center increases as the particle moves farther away, but nucleons can certainly escape from nuclei when given enough energy. However, the last nucleon in a nucleus is typically bound by about 8 MeV, and this has the effect that its wave function drops off rapidly outside of the potential; i.e., the probability of finding it outside the potential is generally small. In fact, the harmonic oscillator potential gives remarkably good wave functions, which can only be improved upon by very detailed calculations.

---

[b]In heavier nuclei such as ^{208}Pb the angular momenta become so large that the spin-orbit splitting is not small compared to the shell spacing, as we shall see.

In the paper of the "Bethe Bible" coauthored with R. F. Bacher,[5] nuclear masses had been measured accurately enough in the vicinity of ^{16}O so that "It may thus be said safely that the completion of the neutron-proton shell at ^{16}O is established beyond doubt from the data about nuclear masses." The telltale signal of the closure of a shell is that the binding energy of the next particle added to the closed shell nucleus is anomalously small. Bethe and Bacher (p. 173) give the shell model levels in the harmonic oscillator potential, the infinite square well, and the finite square well. These figures would work fine for the closed shell nuclei ^{16}O and ^{40}Ca. In Fig. 1, ^{16}O would result from filling the $1s$ and $1p$ shells, ^{40}Ca from filling additionally the $2s$ and $1d$ shells. However, other nuclei that were known to be tightly bound, such as ^{208}Pb, could not be explained by these simple potentials.

The key to the Goeppert Mayer-Jensen success was the spin-orbit splitting, as it turned out. There were the "magic numbers," the large binding energies of ^{16}O, ^{40}Ca, and ^{208}Pb. Especially lead, with 82 protons and 126 neutrons, is very tightly bound. Now it turns out (see Fig. 3) that with the strong spin-orbit splitting the $1h_{11/2}$ level for protons and the $1i_{13/2}$ level for neutrons lie in both cases well below the next highest levels. The notation here is different from that used in atomic physics. The "1" denotes that this is the first time that an h ($l = 5$) level would be filled in adding protons to the nucleus and for the neutron levels similarly, i denoting $l = 6$.

So how could a model in which nucleons move around in a common potential without hitting each other be reconciled with the previous "nuclear porridge" of Niels Bohr? One explanation is that the thorough mixture of particles in the porridge arose because the neutron was dropped into the nucleus with an energy of around 8 MeV above the ground state energy in the cases of "porridge." In 1958 Landau formulated his theory of Fermi liquids, showing that as a particle (fermion) was added with an energy just above the highest occupied state (just at the top of the Fermi sea), the added particle would travel forever without exciting the other particles. In more physical terms, the mean free path (between collisions) of the particle is proportional to the inverse of the square of the difference of its momentum from that of the Fermi surface; i.e.,

$$\lambda \sim \frac{C}{(k - k_{\rm F})^2}, \qquad (13)$$

where C is a constant that depends on the interaction. Thus, as $k \to k_{\rm F}$, $\lambda \to \infty$ and the particle never scatters. Here $k_{\rm F}$, the Fermi momentum, is that of the last filled orbit.

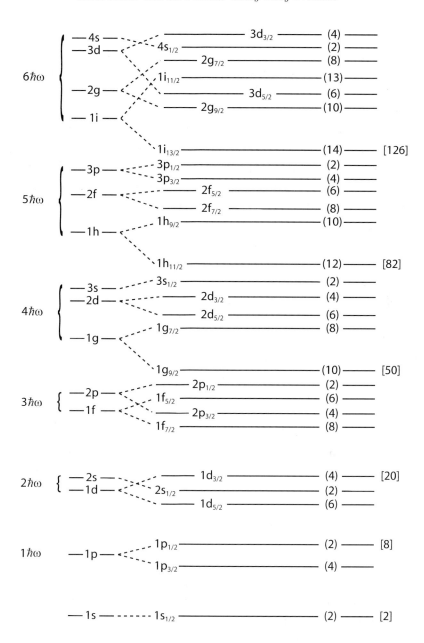

Fig. 3 Sequence of single particle states in the nuclear shell model.

Once the surprise had passed that one could assign a definite shell model state to each nucleon and that these particles moved rather freely, colliding relatively seldomly, the obvious question was how the self-consistent potential the particles moved in could be constructed from the interactions between the particles. The main technical problem was that these forces were very strong. Indeed, Jastrow[6] characterized the short range force between two nucleons as a vertical hard core of infinite height and radius of 0.6×10^{-13} cm, about one-third the average distance between two nucleons in nuclei. (Later, the theoretical radius of the core shrank to 0.4×10^{-13} cm.) The core was much later found to be a rough characterization of the short-range repulsion from vector meson exchange.

The whole postwar development of quantum electrodynamics by Feynman and Schwinger was in perturbation theory, with expansion in the small parameter $\alpha = e^2/\hbar c = 1/137$. Of course, infinities were encountered, but since they shouldn't be there, they were set to zero. The concept of a hard core potential of finite range, the radius a reasonable fraction of the average distance between particles, was new. Lippmann and Schwinger[7] had already developed a formalism that could deal with such an interaction. In perturbation theory; i.e., expansion of an interaction which is weak relative to other quantities, the correction to the energy resulting from the interaction is obtained by integrating the perturbing potential between the wave functions; i.e.,

$$\Delta E \simeq \int \psi_0^\dagger(x,y,z)\, V(x,y,z)\, \psi_0(x,y,z)\, dx\, dy\, dz. \qquad (14)$$

Here $\psi_0(x,y,z)$ is the solution of the Schrödinger wave equation for the zero-order problem, that with $V(x,y,z) = 0$, and $\psi_0^\dagger$ is the complex conjugate of ψ_0. The wavefunction gives the probability of finding a particle located in a small region $dx\, dy\, dz$ about the point (x,y,z):

$$P(x,y,z) = \psi_0^\dagger(x,y,z)\, \psi_0(x,y,z)\, dx\, dy\, dz. \qquad (15)$$

The integral in Eq. (14) is carried out over the entire region in which $V(x,y,z)$ is nonzero.

We see the difficulty that arises if V is infinite, as in the hard core potential; namely, the product of an infinite V and finite ψ is infinite. Equation (14) is just the first-order correction to the energy. More generally, perturbation theory yields a systematic expansion for the energy shift in terms of integrals involving higher powers of the interaction ($V^2, V^3, \cdots$). As long as $\psi(x,y,z)$ is nonzero, all of these terms are infinite.

Watson[8,9] realized that a repulsion of any strength, even an infinitely high hard core, could be handled by the formalism of Lippmann and Schwinger, which was invented in order to handle two-body scattering. Quantum mechanical scattering is described by the T-matrix:

$$T = V + V\frac{1}{E - H_0}V + V\frac{1}{E - H_0}V\frac{1}{E - H_0}V + \cdots, \qquad (16)$$

where V is the two-body potential, E is the unperturbed energy, and H_0 is the unperturbed Hamiltonian, containing only the kinetic energy. This infinite number of terms could be summed to give the result

$$T = V + V\frac{1}{E - H_0}T. \qquad (17)$$

One can see that this is true by rewriting Eq. (16) as

$$T = V + V\frac{1}{E - H_0}\left(V + V\frac{1}{E - H_0}V + V\frac{1}{E - H_0}V\frac{1}{E - H_0}V + \cdots\right), \qquad (18)$$

where the term in parentheses is clearly just T. Watson realized that the T-matrix made sense also for an extremely strong repulsive interaction in a system of many nucleons. An incoming particle could be scattered off each of the nucleons, one by one, and the scattering amplitudes could be added up, the struck nucleon being left in the same state as it was initially. The sum of the amplitudes could be squared to give the total amplitude for the scattering off the nucleus.

Keith Brueckner, who was a colleague of Watson's at the University of Indiana at the time, saw the usefulness of this technique for the nuclear many-body problem. He was the first to recognize that the strong short-range interactions, such as the infinite hard core, would scatter two nucleons to momenta well above those filled in the Fermi sea. Thus, the exclusion principle would have little effect and could be treated as a relatively small correction.

Basically, for the many-body problem, the T-matrix is called the G-matrix, and the latter obeys the equation

$$G = V + V\frac{Q}{e}G. \qquad (19)$$

Whereas in the Lippmann-Schwinger formula Eq. (17) the energy denominator e was taken to be $E - H_0$, there was considerable debate about what to put in for e in Eq. (19). We will see later that it is most conveniently chosen to be $E - H_0$, as in Eq. (17).[c] In Eq. (19), the operator Q excludes from

---

[c]This conclusion will be reached only after many developments, as we outline in Sec. 3.

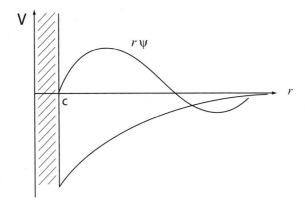

Fig. 4 The wave function ψ for the relative motion of two particles interacting via a potential with an infinite hard core of radius c and an attractive outside potential. It is convenient to deal with $r\psi$ rather than ψ.

the intermediate states not only all of the occupied states below the Fermi momentum k_F, because of the Pauli principle, but also all states beyond a maximum momentum k_{max}. The states below k_{max} define what is called the "model space," which can generally be chosen at the convenience of the investigator. The basic idea is that the solution Eq. (19) of G is to be used as an effective interaction in the space spanned by the Fermi momentum k_F and k_{max}. This effective interaction G is then to be diagonalized within the model space.

In the above discussion leading to Eq. (19) we have written the nucleon-nucleon interaction as V. In fact, the interaction is complicated, involving various combinations of spins and angular momenta of the two interacting nucleons. In the time of Brueckner and the origin of his theory, it took a lot of time and energy just to keep these combinations straight. This large amount of bookkeeping is handled today fairly easily with electronic computers. The major problem, however, was how to handle the strong short-range repulsion, and this problem was discussed in terms of the relatively simple, what we call "central," interaction shown in Fig. 4. In fact, the G of Eq. (19) is still calculated from that equation today in the most successful effective nuclear forces. The k_{max} is taken to be the maximum momentum at which experiments have been analyzed, $k_{max} = 2.1 \text{ fm}^{-1} = \Lambda$, where Λ is now interpreted as a cutoff. Since experiments of momenta higher than Λ have not been carried out and analyzed, at least not in such a way as to bear directly on the determination of the potential, one approach is to leave them out completely. The only important change since the 1950's is that the $V(r)$, the first term on the right-hand side of Eq. (19), is now rewritten in terms of a sum over momenta, by what is called a Fourier transform, and this sum is truncated at Λ, the higher momenta being discarded. The resulting effective interaction, which replaces G, is now called $V_{\text{low}-k}$.[10,11] We expand on this discussion at the end of this article.

The G-matrix of Brueckner was viewed by nuclear physicists as a complicated object. However, it is clear what the effect of a repulsion at core radius c that rises to infinite height will be — it will stop the two interacting particles from going into the region of $r < c$. In non-relativistic quantum mechanics, this means that their wave function of relative motion must be zero. In other words, the wave function, whose square gives the probability of finding the particle in a given region, must be zero inside the hard core. Also, the wave function must be continuous outside, so that it must start from zero at $r = c$. Therefore, we know that the wave function must look something like that shown in Fig. 4. In any case, given the boundary condition $\psi = 0$ at $r = c$, and the potential energy V, the wave equation can be solved for $r > c$. It is not clear at this point how $V(r)$ is to be determined and what the quantity ψ is. We put off further discussion of this until Sec. 3, in which we develop Hans Bethe's "Reference Spectrum."

One of the most important, if not the most important, influences on Hans Bethe in his efforts to give a basis for the nuclear shell model was the work of Feshbach, Porter, and Weisskopf.[12] These authors showed that although the resonances formed by neutrons scattered by nuclei were indeed very narrow, their strength function followed the envelope of a single-particle potential; i.e., of the strength function for a single neutron in a potential $V + iW$. The strength function for the compound nucleus resonances is defined as $\bar{\Gamma}_n/D$, where Γ_n is the width of the resonance for elastic neutron scattering (scattering without energy loss) and D is the average spacing between resonances. This function gives the strength of absorption, averaged over many of the resonances. Parameters of the one-body potential are given in the caption to Fig. 5. The parameter a is the surface thickness of the one-body potential.

The peaks in the neutron strength function occur at those mass numbers where the radius of the single-particle potential is just big enough to bind another single-particle state with zero angular momentum. Although the single-particle resonance is split up into the many narrow states discussed by Bohr, a vestige of the single-particle shell model resonance still remains.

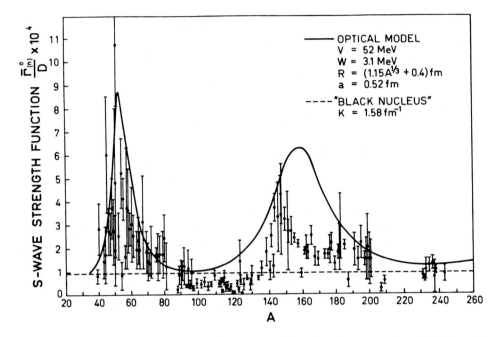

Fig. 5 The S-wave neutron strength function as a function of mass number. This figure is taken from A. Bohr and B. R. Mottelson, *Nuclear Structure*, Vol. 1, p. 230, W. A. Benjamin, New York, 1969.

By contrast, an earlier literal interpretation of Bohr's model was worked out by Feshbach, Peaslee, and Weisskopf[13] in which the neutron is simply absorbed as it enters the nucleus. This curve is the dashed line called "Black Nucleus" in Fig. 5. It has no structure and clearly does not describe the variations in the averaged neutron strength function. The Feshbach, Porter, and Weisskopf paper was extremely important in showing that there was an underlying single-particle shell model structure in the individually extremely complicated neutron resonances.

In fact, the weighting function $\rho(E - E')$ used by Feshbach, Porter, and Weisskopf[12] to calculate the average of the scattering amplitude $S(E)$

$$\langle S(E) \rangle_{Av} = \int_{-\infty}^{\infty} \rho(E - E') F(E') dE' , \qquad (20)$$

where $F(E)$ is an arbitrary (rapidly varying) function of energy E, was a square one that had end effects which needed to be thrown away. A much more elegant procedure was suggested by Jim Langer (see Brown[14]), which involved using the weighting function

$$\rho(E - E') = \frac{I}{\pi} \frac{1}{(E - E')^2 + I^2}. \tag{21}$$

With this weighting function, the average scattering amplitude was

$$\left\{ \sum_n \frac{\Gamma(n)}{W_n - E} \right\}_{Av} = \sum_n \frac{\Gamma(n)}{W_n - E - iI}, \tag{22}$$

where W_n are the energies of the compound states with eV widths. Since the W_n are complex numbers all lying in the lower half of the complex plane, the evaluation of the integral is carried out by contour integration, closing the contour about the upper half plane.

Now if the I is chosen to be about equal to the widths of the single-neutron states in the optical model $V + iW$; i.e., of the order of $W = 3.1$ MeV, then the imaginary part of this average can be obtained as

$$\text{Im}\left(\left\{\sum_n \frac{\Gamma(n)}{W_n - E}\right\}_{Av}\right) = \pi \frac{\bar{\Gamma}(n)}{D} = \text{Im} \sum_m \left(\frac{\Gamma_m}{\hat{E}_m - E}\right), \tag{23}$$

where Γ_m and $\hat{E}_m$ are the widths and (complex) energies of the single neutron states in the complex potential $V + iW$. This shows how the averaged strength function is reproduced by the single-particle levels.

In the summer of 1958 Hans Bethe invited me (G.E.B.) to Cornell, giving me an honorarium from a fund that the AVCO Company, for which he consulted on the physics of the nose cones of rockets upon reentry into the atmosphere, had given him for that purpose. I was unsure of the convergence of the procedure by which I had obtained the above results. Hans pointed out that the width of the single-neutron state $\text{Im}(\Gamma_m)$ would be substantially larger than the widths of the two-particle, one-hole states that the single-particle state would decay into, because in the latter the energy would be divided into the three excitations, two particles and one hole, and the widths went quadratically with available energy as can be obtained from Eq. (13); ergo the two-particle, one-hole widths would be down by a factor of $\sim 3/9 = 1/3$ from the single-particle width, but nonetheless would acquire the same imaginary part I, which is the order of the single particle width in the averaging. Thus, my procedure would be convergent. So I wrote my *Reviews of Modern Physics* article,[14] which I believe was quite elegant, beginning with Jim Langer's idea about averaging and ending with Hans Bethe's argument about convergence. I saw then clearly the advantage of having good collaborators, but had to wait two more decades until I could lure Hans back into astrophysics where we would really collaborate tightly.

2. Hans Bethe in Cambridge, England

We pick up Hans Bethe in 1955 at the time of his sabbatical in Cambridge, England. The family, wife Rose, and children, Monica and Henry, were with him.

There was little doubt that Keith Brueckner had a promising approach to attack the nuclear many-body problem; i.e., to describe the interactions between nucleons in such a way that they could be collected into a general self-consistent potential. That potential would have its conceptual basis in the Hartree-Fock potential and would turn out to be the shell model potential, Hans realized.

Douglas Hartree was a fellow at St. John's College when he invented the self-consistent Hartree fields for atoms in 1928. He put Z electrons into wave functions around a nucleus of Z protons, $A-Z$ neutrons, the latter having no effect because they had no charge. The heavy compact nucleus was taken to be a point charge at the origin of the coordinate system, because the nucleon mass is nearly 2000 times greater than the electron and the size of the nucleus is about 10,000 times smaller in radius than that of the atom. The nuclear charge Ze is, of course, screened by the electron charge as electrons gather about it. The two innermost electrons are very accurately in $1s$ orbits (called K-electrons in the historical nomenclature). Thus, the other $Z-2$ electrons see a screened charge of $(Z-2)e$, and so it could go, but Hartree used instead the so-called Thomas-Fermi method to get a beginning approximation to the screened electric field.

Given this screened field as a function of distance r, measured from the nucleus located at $r = 0$, Hartree then sat down with his mechanical computer punching buttons as his Monromatic or similar machine rolled back and forth, the latter as he hit a return key. (G.E.B. — These machines tore at my eyes, giving me headaches, so I returned to analytical work in my thesis in 1950. Hans used a slide rule, whipping back and forth faster than the Monromatic could travel, achieving three-figure accuracy.) When Hartree had completed the solution of the Schrödinger equation for each of the original Z electrons, he took its wave function and squared it. This gave him the probability, $\rho_i(x_i, y_i, z_i)$, of finding electron i at the position given by the coordinates x_i, y_i, z_i. Then, summing over i, with $x_i = x$, $y_i = y$, and $z_i = z$

$$\sum_i \rho_i(x,y,z) = \rho_e^{(1)}(x,y,z) \qquad (24)$$

gave him the total electron density at position (x, y, z). Then he began over again with a new potential

$$V^{(1)} = \frac{Ze}{r} - e\rho_e^{(1)}(x,y,z), \qquad (25)$$

where superscript (1) denotes that this is a first approximation to the self-consistent potential. To reach approximation (2) he repeated the process, calculating the Z electronic wave functions by solving the Schrödinger equation with the potential $V^{(1)}$. This gave him the next Hartree potential $V^{(2)}$. He kept going until the potential no longer changed upon iteration; i.e., until $V^{(n+1)} \simeq V^{(n)}$, the $\simeq$ meaning that they were approximately equal, to the accuracy Hartree desired. Such a potential is called "self-consistent" because it yields an electron density that reproduces the same potential.

Of course, this was a tedious job, taking months for each atom (now only seconds with electronic computers). Some of the papers are coauthored, D. R. Hartree and W. Hartree. The latter Hartree was his father, who wanted to continue working after he retired from employment in a bank. In one case, Hartree made a mistake in transforming his units Z and e to more convenient dimensionless units, and he performed calculations with these slightly incorrect units for some months. Nowadays, young investigators might nonetheless try to publish the results as referring to a fractional Z, hoping that fractionally charged particles would attach themselves to nuclei, but Hartree threw the papers in the wastebasket and started over.

Douglas Hartree was professor in Cambridge in 1955 when Hans Bethe went there to spend his sabbatical. The Hartree method was improved upon by the Russian professor Fock, who added the so-called exchange interaction which enforced the Pauli exclusion principle, guaranteeing that two electrons could not occupy the same state. We shall simply enforce this principle by hand in the following discussion, our purpose here being to explain what "self-consistent" means in the many-body context. Physicists generally believe self-consistency to be a good attribute of a theory, but Hartree did not have to base his work only on beliefs. Given his wave functions, a myriad of transitions between atomic levels could be calculated, and their energies and probabilities could be compared with experiment. Douglas Hartree became a professor at Cambridge. This indicates the regard in which his work was held.

The shell model for electrons reached success in the Hartree-Fock self-consistent field approach. This was very much in Hans Bethe's mind, when he set out to formulate Brueckner theory so that he could obtain a self-consistent potential for nuclear physics. He begins his 1956 paper[15] on the "Nuclear Many-Body Problem" with "Nearly everybody in nuclear physics has marveled at the success of the shell model. We shall use the expression 'shell model' in its most general sense, namely as a scheme in which each nucleon is given its individual quantum state, and the nucleus as a whole is described by a 'configuration,' i.e., by a set of quantum numbers for the individual nucleons."

He goes on to note that even though Niels Bohr had shown that low-energy neutrons disappear into a "porridge" for a very long time before re-emerging (although Hans was not influenced by Bohr's paper, because he hadn't read it; this may have given him an advantage as we shall see), Feshbach, Porter, and Weisskopf had shown that the envelope of these states followed that of the single particle state calculated in the nuclear shell model, as we noted in the last section.

Bethe confirms "while the success of the model [in nuclear physics] has thus been beyond question for many years, a theoretical basis for it has been lacking. Indeed, it is well established that the forces between two nucleons are of short range, and of very great strength, and possess exchange character and probably repulsive cores. It has been very difficult to see how such forces could lead to any over-all potential and thus to well-defined states for the individual nucleons."

He goes on to say that Brueckner has developed a powerful mathematical method for calculating the nuclear energy levels using a self-consistent field method, even though the forces are of short range.

"In spite of its apparent great accomplishments, the theory of Brueckner *et al.* has not been readily accepted by nuclear physicists. This is in large measure the result of the very formal nature of the central proof of the theory. In addition, the definitions of the various concepts used in the theory are not always clear. Two important concepts in the theory are the wave functions of the individual particles, and the potential V_c 'diagonal' in these states. The paper by Brueckner and Levinson[16] defines rather clearly how the potential is to be obtained from the wave functions, but not how the wave functions can be constructed from the potential V_c. Apparently, BL assume tacitly that the nucleon wave functions are plane waves, but in this case, the method is only applicable to an infinite nucleus. For a finite nucleus, no prescription is given for obtaining the wave functions."

Hans then goes on to define his objective, which will turn out to develop into his main activity for the next decade or more: "It is the purpose of the present paper to show that the theory of Brueckner gives indeed the foundation of the shell model."

Hans was rightly very complimentary to Brueckner, who had "tamed" the extremely strong short-ranged interactions between two nucleons, often taken to be infinite in repulsion at short distances at the time. On the other hand, Brueckner's immense flurry of activity, changing and improving on previous papers, made it difficult to follow his work. Also, the Watson input scattering theory seemed to give endless products of scattering operators, each appearing to be ugly mathematically ("Taming" thus was the great accomplishment of Watson and Brueckner.) And in the end, the real goal was to provide a quantitative basis for the nuclear shell model, based on the nucleon-nucleon interaction, which was being reconstructed from nucleon-nucleon scattering experiments at the time.

The master at organization and communication took over, as he had done in formulating the "Bethe Bible" during the 1930's.

One can read from the Hans Bethe archives at Cornell that Hans first made 100 pages of calculations reproducing the many results of Brueckner and collaborators, before he began numbering his own pages as he worked out the nuclear many-body problem. He carried out the calculations chiefly analytically, often using mathematical functions, especially spherical Bessel functions, which he had learned to use while with Sommerfeld. When necessary, he got out his slide rule to make numerical calculations.

During his sabbatical year, Hans gave lectures in Cambridge on the nuclear many-body problem. Two visiting American graduate students asked most of the questions, the British students being reticent and rather shy. But Professor Nevill Mott asked Hans to take over the direction of two graduate students, Jeffrey Goldstone and David Thouless, the former now professor at M.I.T. and the latter professor at the University of Washington in Seattle. We shall return to them later.

As noted earlier, the nuclear shell model has to find its own center "self-consistently." Since it is spherically symmetrical, this normally causes no problem. Simplest is to assume the answer: begin with a deep square well or harmonic oscillator potential and fill it with single particle eigenstates as a zero-order approximation as Bethe and Bacher[5] did in their 1936 paper of the Bethe Bible. Indeed, now that the sizes and shapes of nuclei have been measured by high resolution electron microscopes (high energy electron scattering), one can reconstruct these one-particle potential wells so that filling them with particles reproduces these sizes and shapes. The wave functions in such wells are often used as assumed solutions to the self-consistent potential that would be obtained by solving the nuclear many-body problem.

We'd like to give the flavor of the work at the stage of the Brueckner theory, although the work using the "Reference-Spectrum" by Bethe, Brandow, and Petschek,[17] which we will discuss later, will be much more convenient for understanding the nuclear shell model. The question we consider is the magnitude of the three-body cluster terms; i.e., the contribution to the energy from the interaction of three particles. (Even though the elementary interaction may be only a two-particle one, an effective three-body interaction arises inside the nucleus, as we shall discuss.) One particular three-body cluster will be shown in Fig. 10. The three-body term, called a three-body cluster in the Brueckner expansion, is

$$\Delta E_3 = \sum_{ijk} \left\langle \Phi_0 \left| I_{ij} \frac{Q}{e_{ji}} I_{jk} \frac{Q}{e_{ki}} I_{ki} \right| \Phi_0 \right\rangle. \tag{26}$$

Here the three nucleons i, j, and k are successively excited. There are, of course, higher order terms in which more nucleons are successively excited and de-excited. In this cluster term, working from right to left, first there is an interaction I_{ki} between particles i and k. In fact, this pair of particles is allowed to interact any number of times, the number being summed into the G-matrix G_{ik}. We shall discuss the G-matrix in great detail later. Then the operator Q excludes all intermediate states that are occupied by other particles — the particle k can only go into an unoccupied state before interacting with particle j. The denominator e_{ki} is the difference in energy between the specific state k that particle i goes to and the state that it came from. The particle k can make virtual transitions, transitions that don't conserve energy, because of the Heisenberg uncertainty principle

$$\Delta E \Delta t \gtrsim \hbar. \tag{27}$$

Thus, if $\Delta E \Delta t > \hbar$, then the particle can stay in a given state only at time

$$t = \hbar/e_{ki} \tag{28}$$

so the larger the e_{ki} the shorter the time that the particle in a given state can contribute to the energy ΔE_3. (Of course, the derivation of ΔE_3 is carried out in the standard operations of quantum mechanics. We bring in the uncertainty principle only to give some qualitative understanding of the result.) Once particle k has interacted with particle j, particle j goes on to interact with the original particle i, since the nucleus must be left in the same ground state Φ_0 that it began in, if the three-body cluster is to contribute to its energy.

Bethe's calculation gave

$$\Delta E_3/A = -0.66 \text{ MeV} \tag{29}$$

corrected a bit later in the paper to -0.12 MeV once only the fraction of spin-charge states allowed by selection rules are included. This is to be compared with Brueckner's -0.007 MeV. Of course, these are considerably different, but this is not the main point. The main point is that both estimates are small compared with the empirical nuclear binding energy.[d]

Thus, it was clear that the binding energy came almost completely from the two-body term G_{ik}, and that the future effort should go into evaluating this quantity, which satisfies the equation

$$G = V + V\frac{Q}{e}G. \tag{30}$$

(We will have different G's for the different charge-spin states.)

Although written in a deceptively simple way, this equation is ugly, involving operators in both the coordinates x, y, z, and their derivatives. However, G can be expressed as a two-body operator (see Sec. 3). It does not involve a sum over the other particles in the nucleus, so the interactions can be evaluated one pair at a time.

Thus, the first paper of Bethe on the nuclear many-body problem collected the work of Brueckner and collaborators into an orderly formalism in which the evaluation of the two-body operators G would form the basis for calculating the shell model potential $V(r)$. Bethe went on with Jeffrey Goldstone to investigate the evaluation of G for the extreme infinite-height hard core potential, and he gave David Thouless the problem that, given the empirically known shell model potential $V_{SM}(r)$, what properties of G would reproduce it.

As we noted in the last section, an infinite sum over the two-body interaction is needed in order to completely exclude the two-nucleon wave functions from the region inside the (vertical) hard core. As noted earlier,

---

[d] At least at that time there appeared to be a good convergence in the so-called cluster expansion. But see Sec. 3!

Jastrow[6] had proposed a vertical hard core, rising to ∞, initially of radius $R_c = 0.6 \times 10^{-13}$ cm. This is about 1/3 of the distance between nucleons, so that removing this amount of space from their possible occupancy obviously increases their energy. (From the Heisenberg principle $\Delta p_x \Delta x \gtrsim \hbar$, which is commonly used to show that if the particles are confined to a smaller amount of volume, the Δx is decreased, so the Δp_x, which is of the same general size as p_x, is increased.) Thus, as particles were pushed closer together with increasing density, their energy would be greater. Therefore, the repulsive core was thought to be a great help in saturating matter made up of nucleons. One of the authors (G.E.B.) heard a seminar by Jastrow at Yale in 1949, and Gregory Breit, who was the leading theorist there, thought well of the idea. Indeed, we shall see later that Breit played an important role in providing a physical mechanism for the hard core as we shall discuss in the next section. This mechanism actually removed the "sharp edges" which gave the vertical hard core relatively extreme properties. So we shall see later that the central hard core in the interaction between two nucleons isn't vertical, rather it's of the form

$$\text{hard core} = \frac{C}{r} \exp\left(-\frac{m_\omega c r}{\hbar}\right), \tag{31}$$

where $\hbar/m_\omega c$ is 0.25×10^{-13} cm, and m_ω is the mass of the ω-meson. The constant C is large compared with unity

$$C \gg 1 \tag{32}$$

so that the height of the core is many times the Fermi energy; i.e., the energy measured from the bottom of the shell model potential well up to the last filled level. Thus, although extreme, treating the hard core potential gives a caricature problem. Furthermore, solving this problem in "Effect of a repulsive core in the theory of complex nuclei," by H. A. Bethe and J. Goldstone[18] gave an excellent training to Jeffrey Goldstone, who went on to even greater accomplishments. (The Goldstone boson is named after him; it is perhaps the most essential particle in QCD.)

We will find in the next section that the Moszkowski-Scott separation method is a more convenient way to treat the hard core. And this method is more practical because it includes the external attractive potential at the same time. However, even though the wave function of relative motion of the two interacting particles cannot penetrate the hard core, it nonetheless has an effect on their wave function in the external region. This effect is conveniently included by changing the reduced mass of the interacting nucleons from $M^\star$ to $0.85 M^\star$ for a hard core radius $r_c = 0.5 \times 10^{-13}$ cm, and to $\sim 0.9 M^\star$ for the presently accepted $r_c = 0.4 \times 10^{-13}$ cm.

3. The Reference-Spectrum Method for Nuclear Matter

We take the authors' prerogative to jump in history to 1963, to the paper of H. A. Bethe, B. H. Brandow, and A. G. Petschek[17] with the title of that of this section, because this work gives a convenient way of discussing essentially all of the physical effects found by Bethe and collaborators, and also the paper by S. A. Moszkowski and B. L. Scott[19] the latter paper being very important for a simple understanding of the G-matrix. In any case, the reference-spectrum included in a straightforward way all of the many-body effects experienced by the two interacting particles in Brueckner theory, and enabled the resulting G-matrix to be written as a function of r alone, although separate functions $G_l(r)$ had to be obtained for each angular momentum l.

Let us prepare the ground for the reference-spectrum, including also the Moszkowski-Scott method, by a qualitative discussion of the physics involved in the nucleon-nucleon interaction inside the nucleus, say at some reasonable fraction of nuclear matter density. We introduce the latter so we can talk about plane waves locally, as an approximation. For simplicity we take the short-range repulsion to come from a hard core, of radius 0.4×10^{-13} cm.

(i) In the region just outside the hard core, the influence of neighboring nucleons on the two interacting ones is negligible, because the latter have been kicked up to high momentum states by the strong hard-core interaction and that of the attractive potential, which is substantially stronger than the local Fermi energy of the nucleons around the two interacting ones. Thus, one can begin integrating out the Schrödinger equation for $r\psi(r)$, starting from $r\psi(r) = 0$ at $r = r_c$. The particles cannot penetrate the infinitely repulsive hard core (see Fig. 6), so their wave function begins from zero there. In fact, the Schrödinger equation is one of relative motion; i.e., a one-body equation in which the mass is the reduced mass of the two particles, $M_N/2$ for equal mass nucleons. As found by Bethe and Goldstone, the M_N will be changed to $M_N^\star$ in the many-body medium, but the mass is modified not only by the hard core but also by the attractive part of the two-body potential, which we have not yet discussed.

(ii) We consider the spin-zero $S = 0$ and spin-one $S = 1$ states; i.e., for angular momentum $l = 0$. These are the most important states. We compare

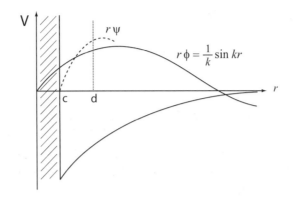

Fig. 6 The unperturbed S-wave function $r\phi = \frac{1}{k}\sin kr$ and the wave function $r\psi$ obtained by integrating the Schrödinger equation in the presence of the potential out from $r = c$.

the spin-one state in the presence of the potential with the unperturbed one; i.e., the one in the absence of a potential.

Before proceeding further with Hans Bethe's work, let us characterize the nice idea of Moszkowski and Scott in the simplest possible way.

Choose the separation point d such that

$$\left.\frac{d\psi(r)/dr}{\psi(r)}\right|_{r=d} = \left.\frac{d\phi(r)/dr}{\phi(r)}\right|_{r=d} \tag{33}$$

Technically, this is called the equality of logarithmic derivatives. As shown in Fig. 6, such a point will be there, because $r\psi$, although it starts from zero at $r = c$, has a greater curvature than $r\phi$. Now if only the potential inside of d were present; i.e., if $V(r) = 0$ for $r > d$, then Eq. (33) in quantum mechanics is just the condition that the inner potential would produce zero scattering, the inner attractive potential for $r < d$ just canceling the repulsion from the hard core. Of course, if Eq. (33) is true for a particular k, say $k \lesssim k_F$, then it will not be exactly true for other k's. On the other hand, since the momenta in the short-distance wave function are high compared with k_F, due to the infinite hard core, and the very deep interior part of the attractive potential that is needed to compensate for it, Eq. (33) is nearly satisfied for all momenta up to k_F if the equality is true for one of the momenta.

The philosophy here is very much as in Bethe's work "Theory of the Effective Range in Nuclear Scattering."[20] This work is based on the fact that the inner potential (we could define it as $V(r)$ for $r < d$) is deep in comparison with the energy that the nucleon comes in with, so that the scattering depends only weakly on this (asymptotic) energy. In fact, for a potential which is Yukawa in nature

$$V = \infty, \qquad \text{for } r < c = 0.4 \times 10^{-13} \text{ cm} \qquad (34)$$

$$V = -V_0 e^{-(r-c)/R} \quad \text{for } r > c, \qquad (35)$$

then $V_0 = 380$ MeV and $R = 0.45 \times 10^{-13}$ cm in order to fit the low energy neutron-proton scattering.[21] The value of 380 MeV is large compared with the Fermi energy of ~ 40 MeV. In the collision of two nucleons, the equality of logarithmic derivatives, Eq. (33), would mean that the inner part of the potential interaction up to d, which we call V_s, would give zero scattering. All the scattering would be given by the long-range part which we call V_l.

(iii) Now we know that the wave function $\psi(r)$ must "heal" to $\phi(r)$ as $r \to \infty$, because of the Pauli principle. There is no other place for the particle to go, because for k below k_F all other states are occupied. Delightfully simple is to approximate the healing by taking ψ equal to ϕ for $r > d$.

The conclusion of the above is that

$$G(r) \simeq G_s + v_l(r) + v_l \frac{Q}{e} v_l + \cdots, \qquad (36)$$

where

$$v_l = V(r) \quad \text{for } r \gtrsim d. \qquad (37)$$

We shall discuss G_s, the G-matrix which would come from the short-range part of the potential for $r < d$ later. It will turn out to be small.

Now we have swept a large number of problems under the rug, and we don't apologize for it because Eq. (36) gives a remarkably accurate answer. However, first we note that there must be substantial attraction in the channel considered in order for the short-range repulsion to be canceled. So the separation method won't work for cases where there is little attraction. Secondly, a number of many-body effects have been discarded, and these had to be considered by Bethe, Brandow, and Petschek, one by one.

We do not want to fight old battles over again, but simply note that Bethe, Brandow, and Petschek found that they could take into account the many-body effects including the Pauli principle by choosing a single-particle spectrum, which is approximated by

$$\epsilon_m = \frac{k_m^2}{2m_{RS}^\star} + A. \qquad (38)$$

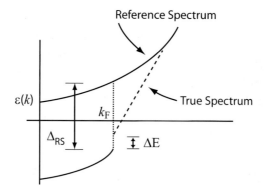

Fig. 7 Energy spectrum in the Brueckner theory (inclusive of self-energy insertion). The true spectrum has a small gap ΔE at k_F and is known, from the arguments of Bethe et al., to join the reference-spectrum at $k \approx 3$ fm^{-1}. The dotted line is an interpolation between these two points.

The hole energies, i.e., the energies of the initially bound particles, can be treated fairly roughly since they are much smaller than the energies of the particles which have quite high momenta (see Fig. 7). The parameters $m^\star_{RS}$ and A were arrived at by a self-consistency process. Given an input $m^\star_{RS}$ and A, the same $m^\star_{RS}$ and A must emerge from the calculation. Thus, the many-body effects can be included by changing the single-particle energy through the coefficient of the kinetic energy and by adding the constant A. In this way the many-body problem is reduced to the Lippmann-Schwinger two-body problem with changed parameters, $m \to m^\star_{RS}$ and the addition of A. As shown in Fig. 7, the effect is to introduce a gap Δ_{RS} between particle and hole states.[e]

A small gap, shown as ΔE, occurs naturally between particle and hole states, resulting from the unsymmetrical way they are treated in Brueckner theory, the particle-particle scattering being summed to all orders. The main part of the reference-spectrum gap Δ_{RS} is introduced so as to numerically reproduce the effects of the Pauli Principle. That this can be done is not surprising, since the effect of either is to make the wave function heal more rapidly to the noninteracting one.

Before we move onwards from the reference-spectrum, we want to show the main origin of the reduction of m to $m^\star_{RS} \sim 0.8m$. Such a reduction obviously increases the energies of the particles in intermediate states $\epsilon(k)$

---

[e]We call the states initially in the Fermi sea "hole states" because holes are formed when the two-body interaction transfers them to the particle states which lie above the Fermi energy, leaving holes behind in intermediate states.

which depends inversely on $m^\star_{RS}$. A lowering of $m^\star$ from m was already found in the Bethe-Goldstone solution of the scattering from a hard core alone with $m^\star = 0.85m$ for a hard core radius of $r_c = 0.5 \times 10^{-13}$ cm and somewhat less for $r_c = 0.4 \times 10^{-13}$ cm, a more reasonable value. The $m^\star_{RS}$ are not much smaller than this.

Now we have to introduce the concept of off-energy-shell self energies. Let's begin by defining the on-shell self-energy which is just the energy of a particle in the shell model or optical model potential. It would be given by the process shown in Fig. 8.

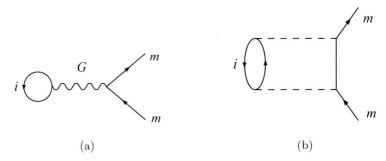

Fig. 8 (a) The on-shell self energy involves the interaction of a particle in state m with the filled states in the Fermi sea i. The wiggly line G implies a sum over all orders in a ladder of V, the second-order term in dashed lines being shown in (b). Together with the kinetic energy, this gives the self energy.

The on-shell $U(k_m)$ should just give the potential energy of a nucleon of momentum k in the optical model potential $U + iW$ used by Feshbach, Porter, and Weisskopf, so $U(k_m) + k_m^2/2m$ gives the single-particle energy.

However, the energies $U(k_m)$ to be used in Brueckner theory are not on-shell energies. This is because another particle must be excited when the one being considered is excited, as shown in Fig. 9. Thus, considering the second

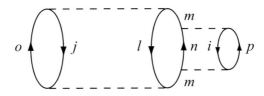

Fig. 9 Second-order contribution to the self energy of a particle in state m when present in a process contributing to the ground-state energy. The states m, n, o, p refer to particles and i, j, l to holes.

order in V interaction between the particle being considered and those in the nucleus one has

$$\tilde{\epsilon}_m^{(2)} = -\sum_{p,n>k_F; i<k_F} \frac{|\langle np|V|im\rangle|^2}{\epsilon_n + \epsilon_p + \epsilon_o - \epsilon_j - \epsilon_i - \epsilon_l} \qquad (39)$$

i.e., although the particle in state o and the hole in state j do not enter actively into the self energy of p, one must include their energies in the energy denominator. Defining

$$\tilde{\epsilon} = \epsilon_n + \epsilon_p + \epsilon_o - \epsilon_j - \epsilon_i - \epsilon_l, \qquad (40)$$

$$e = \epsilon_n + \epsilon_p - \epsilon_i - \epsilon_l \qquad (41)$$

one sees that the additional energy in the denominator in Eq. (39) is

$$\Delta e = \tilde{\epsilon} - \epsilon = \epsilon_o - \epsilon_j. \qquad (42)$$

This Δe represents the energy that must be "borrowed" in order to excite the particle from state j to o, even though this particle does not directly participate in the interaction on state p which gives its self energy. The point is that because j must also be excited, as well as the particles on the right in the diagram of Fig. 9, the total $\tilde{e}$ is that much greater than the e necessary to give the on-shell energy. Thus, the time that this interaction can go on for is decreased, since, from the uncertainty principle the time that energy can be borrowed for goes as $\Delta t = \hbar/\Delta E$. As noted earlier, the wave function $\psi(r)$ lies outside the hard core, so that integrals over the G-matrix are only over the attractive interaction. These are cut down by the additional energy denominator that comes in by the interaction being off shell. In the higher momentum region this can cut the self energy down by 15–20 MeV. We shall see later that when this is handled properly, only $\sim 1/3$ of this survives.

The strategy that T. T. S. Kuo and one of the authors (G.E.B.) have employed over many years, beginning with Kuo and Brown[22] to calculate the G-matrix in nuclei has been to use the separation method for the G in the $l=0$ states (S-states), which have a large enough attraction so that a separation distance d can be defined, but to use the Bethe, Brandow, and Petschek reference-spectrum in angular momentum states which do not have this strong attraction.

The reference-spectrum was as far as one could get using only two-body clusters; i.e., summing up the two-body interaction. In order to make further progress, three and four-body clusters had to be considered; i.e., processes in which one pair of the initially interacting particles was left excited while another pair interacted, and only then returned to its initial state as in the calculations of the lowest order three-body cluster ΔE_3, Eq. (26).

Hans Bethe investigated[23] the four-body clusters, following the work[24] on three-nucleon clusters in nuclear matter by his post-doc R. (Dougy) Rajaraman, a paper just following the Bethe, Brandow, and Petschek reference-spectrum paper. Rajaraman suggested on the basis of including the other three-body clusters than that shown in Fig. 8, that the off-shell effect should be decreased by a factor of $\sim 1/2$. (We shall see from Bethe's work below that it should actually be more like $\sim 1/3$.) Bethe first made a fairly rough calculation of the four-body clusters, but good enough to show when summed to all orders in G, it was tremendous, giving

$$\Delta E_4 \simeq -35 \text{ MeV/particle} \qquad (43)$$

to the binding energy.

Visiting CERN in the summer of 1964, Bethe learned of Faddeev's solution of the three-body problem.[25] This formalism for three-body scattering is very similar to that for two-body scattering in Eq. (17) in that one can define a T-matrix

$$T\Phi = V\psi, \qquad (44)$$

where ψ and Φ are the three-particle correlated and uncorrelated wave functions, and T satisfies

$$T = V + V\frac{1}{e}T \qquad (45)$$

where e now represents the energy denominator of all three particles. The T can be split into

$$T = T^{(1)} + T^{(2)} + T^{(3)} \qquad (46)$$

but now one defines

$$T^{(1)} = V_{23} + V_{23}\frac{1}{e}T, \qquad (47)$$

etc. In other words, $T^{(1)}$ denotes that part of T in which particle "1" did not take part in the last interaction. Hans found that he could write expressions for the three-body problem using the G-matrix as effective interaction; e.g.

$$T^{(1)}\Phi = G_{23}\psi^{(1)}, \qquad (48)$$

where $\psi^{(1)}$ is defined by

$$\psi^{(1)} = \Phi + \frac{1}{e}(T^{(2)} + T^{(3)})\Phi. \qquad (49)$$

The ground state energy is given by

$$E_0 = \langle \Phi | T | \Phi \rangle, \qquad (50)$$

where Φ is the unperturbed (free-particle) wave function. In short, one could use the Brueckner G-matrices as effective interactions in the Faddeev formalism, which then summed the three-body cluster terms to all orders in the solution of these equations.

Hans came to Copenhagen in late summer of 1964, with the intention of solving the Faddeev equations for the three-body system using the reference-spectrum approximation Eq. (38) for the effective two-body interaction. David Thouless, Hans' Ph.D. student in Cambridge, was visiting Nordita (Nordic Institute of Theoretical Atomic Physics) at that time. I (G.E.B.) got David together with Hans the morning after Hans arrived.[f] Hans wrote down the three coupled equations on the blackboard and began to solve them by some methods Dougy Rajaraman had used for summing four-body clusters. David took a look at the three coupled equations with Hans' G-matrix reference-spectrum approximation for the two-body interaction and said "These are just three coupled linear equations. Why don't you solve them analytically?" By late morning Hans had the solution, given in his 1965 paper. The result, which can be read from this paper, is that the off-shell correction to the three-body cluster of Fig. 10 should be cut down by a factor of 1/3. Hans' simple conclusion was "when three nucleons are close

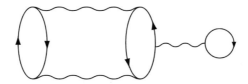

Fig. 10 The contribution of the three-body cluster to the off-shell self energy of a particle in state p. The wiggly line here represents the G-matrix. One particular contribution to this is shown in Fig. 9.

---

[f]Our most fruitful discussions invariably came the morning after he arrived in Copenhagen.

together, an elementary treatment would give us three repulsive pair interactions. In reality, we cannot do more than exclude the wave function from the repulsive region, hence we get only one core interaction rather than 3."

In the summer of 1967, I (G.E.B.) gave the summary talk of the International Nuclear Physics meeting held in Tokyo, Japan. This talk is in the proceedings. I designed my comments as letters to the speakers, as did Herzog in Saul Bellow's novel by that name.

Dear Professor Bethe,

First of all, your note is too short to be intelligible. But by valiant efforts, and a high degree of optimism in putting together corrections of the right sign, you manage to get within 3 MeV/particle of the binding energy of nuclear-matter. It is nice that there are still some discrepancies, because we must have some occupation for theorists, in calculating three-body forces and other effects.

Most significantly, you confirm that it is a good approximation to use plane-wave intermediate states in calculation of the G-matrix. This simplifies life in finite nuclei immensely.

I cannot agree with you that there is no difference between hard- and soft-core potentials. I remember your talk at Paris, where you showed that the so-called dispersion term (the contribution of G_s — G.E.B.), which is a manifestation of off-energy-shell effects, differs by ~ 3 or ~ 4 MeV for hard- and soft-core potentials, and that this should be the only difference. I remain, therefore, a strong advocate of soft-core potentials, and am confident that careful calculation will show them to be significantly better.

Let me remind you that we (Kuo and Brown) always left out this dispersion correction in our matrix elements for finite nuclei, in the hopes of softer ones.

. . . .

Yours, etc.

In summary, the off-energy-shell effects at the time of the reference-spectrum through the dispersion correction G_s contributed about +6 MeV to the binding energy, a sizable fraction of the ~ 15 MeV total binding energy. However, Bethe's solution of the three-body problem via Faddeev cut this by 1/3. The reference-spectrum paper was still using a vertical hard core, whereas the introduction of the Yukawa-type repulsion by Breit[26] and independently by Gupta,[27] although still involving Fourier (momentum) components of $p \sim m_\omega c$ with m_ω the 782 MeV/c^2 mass of the ω-meson, $\sim 2\frac{1}{2}$ times greater than k_F, cut the dispersion correction down somewhat more, so we were talking in 1967 about a remaining ~ 1 MeV compared with the -16 MeV binding energy per particle. At that stage we agreed to neglect interactions in the particle intermediate energy states, as is done in the Schwinger-Dyson Interaction Representation. However, in the latter, any interactions that particles have in intermediate states are expressed in terms of higher-order corrections, whereas we just decided ("decreed") in 1967 that these were negligible. An extensive discussion of all of the corrections to the dispersion term from three-body clusters and other effects is given in Michael Kirson's Cornell 1966 thesis, written under Hans' direction. In detail, Kirson was not quite as optimistic about dropping all off-shell effects as we have been above, but he does find them to be at most a few MeV.

Once it was understood that the short-range repulsion came from the vector meson exchange potentials, rather than a vertical hard core, the short-range G-matrix G_s in Eq. (36) could be evaluated and it was found to be substantially smaller than that for the vertical hard core. Because of the smoothness in the potential, the off-shell effects were smaller, sufficiently so that G_s could be neglected. Thus, the short-range repulsion was completely tamed and one had to deal with only the well-behaved power expansion

$$G = v_l + v_l \frac{Q}{e} v_l + \cdots \tag{51}$$

In fact, in the $^1S_0$ states the first term gave almost all of the attraction, whereas the iteration of the tensor force in the second term basically gave the amount that the effective $^3S_1$ potential exceeded the $^1S_0$ one in magnitude. (The tensor force contributed to only the triplet states because it required $S = 1$.)

Bethe's theory of the effective range in nuclear scattering[20] showed how the scattering length and effective range, the first two terms in the expansion of the scattering amplitude with energy, could be obtained from any potential. The scattering length and effective range could be directly obtained from the experimental measurements of the scattering.

Thus, one knew that any acceptable potential must reproduce these two constants, and furthermore, from our previous discussion, contain a strong short-range repulsion. To satisfy these criteria, Kallio and Kolltveit[28] therefore took singlet and triplet potentials

$$V_i(r) = \begin{cases} \infty & \text{for } r \leq 0.4 \text{ fm} \\ -A_i e^{-\alpha_i(r-0.4)} & \text{for } r > 0.4 \text{ fm} \end{cases} \quad \text{for } i = s, t, \quad (52)$$

where

$$\begin{aligned} A_s &= 330.8 \text{ MeV}, & \alpha_s &= 2.4021 \text{ fm}^{-1} \\ A_t &= 475.0 \text{ MeV}, & \alpha_t &= 2.5214 \text{ fm}^{-1} \end{aligned} \quad (53)$$

By using this potential, Kallio and Kolltveit obtained good fits to the spectra of light nuclei.

Surprising developments, summarized in the *Physics Report* "Model-Independent Low Momentum Nucleon Interaction from Phase Shift Equivalence,"[10] showed how one could define an effective interaction between nucleons by starting from a renormalization group approach in which momenta larger than a cutoff Λ are integrated out of the effective G-matrix interaction. In fact, the data which went into the phase shifts obtained by various groups came from experiments with center of mass energies less than 350 MeV, which corresponds to a particle momentum of 2.1 fm^{-1}. Thus, setting Λ equal to 2.1 fm^{-1} gave an effective interaction $V_{\text{low}-k}$ which included all experimental measurements.

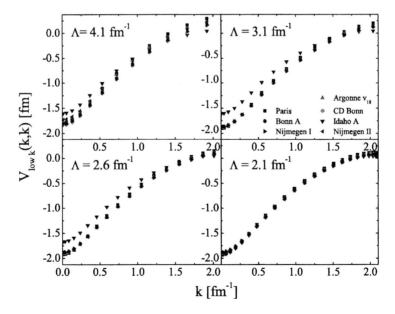

Fig. 11 Diagonal matrix elements of $V_{\text{low}-k}$ for different high-precision potentials in the $^1S_0$ partial wave with various cutoffs Λ.

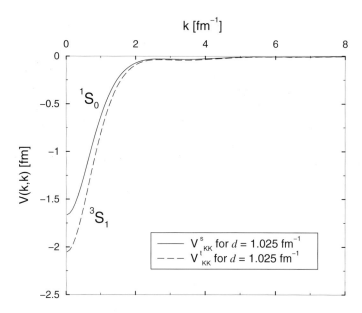

Fig. 15 Diagonal matrix elements of the long-distance Kallio-Kolltveit potential v_l, including momenta above the $V_{\text{low}-k}$ cutoff of 2.1 fm^{-1}.

It is, therefore, clear that the G of Eq. (51) that Bethe ended with is essentially $V_{\text{low}-k}$, the only difference being that he made the separation of scales in configuration space whereas in $V_{\text{low}-k}$ it is made in momentum space. Thus, we believe that Hans Bethe arrived at the right answer in "The Nuclear Many-Body Problem," but only much later did research workers use it to fit spectra.

References

1. N. Bohr, *Nature* **137**, 344 (1936).
2. N. Bohr, *Phil. Mag.* **26**, 476 (1913).
3. M. G. Mayer, *Phys. Rev.* **75**, 1969 (1949).
4. O. Haxel, J. H. D. Jensen and H. E. Suess, *Z. Phys.* **128**, 295 (1950).
5. H. A. Bethe and R. F. Bacher, *Rev. Mod. Phys.* **8**, 82 (1936).
6. R. Jastrow, *Phys. Rev.* **81**, 165 (1951).
7. B. A. Lippmann and J. Schwinger, *Phys. Rev.* **79**, 469 (1950).
8. K. M. Watson, *Phys. Rev.* **89**, 575 (1953).
9. N. C. Francis and K. M. Watson, *Phys. Rev.* **92**, 291 (1953).
10. S. K. Bogner, T. T. S. Kuo and A. Schwenk, *Phys. Rep.* **386**, 1 (2003).
11. G. E. Brown and M. Rho, *Phys. Rep.* **396**, 1 (2004).
12. H. Feshbach, C. E. Porter and V. F. Weisskopf, *Phys. Rev.* **96**, 448 (1954).

13. H. Feshbach, D. C. Peaslee and V. F. Weisskopf, *Phys. Rev.* **71**, 145 (1947).
14. G. E. Brown, *Rev. Mod. Phys.* **31**, 893 (1959).
15. H. A. Bethe, *Phys. Rev.* **103**, 1353 (1956).
16. K. A. Brueckner and C. A. Levinson, *Phys. Rev.* **97**, 1344 (1955).
17. H. A. Bethe, B. H. Brandow and A. G. Petschek, *Phys. Rev.* **129**, 225 (1963).
18. H. A. Bethe and J. Goldstone, *Proc. Roy. Soc. A* **238**, 551 (1957).
19. S. A. Moszkowski and B. L. Scott, *Ann. Phys.* **14**, 107 (1961).
20. H. A. Bethe, *Phys. Rev.* **76**, 38 (1949).
21. G. E. Brown and A. D. Jackson, *The Nucleon-Nucleon Interaction*, North-Holland, Amsterdam, 1976.
22. T. T. S. Kuo and G. E. Brown, *Nucl. Phys.* **85**, 40 (1966).
23. H. A. Bethe, *Phys. Rev.* **138**, B804 (1965).
24. R. Rajaraman, *Phys. Rev.* **129**, 265 (1963).
25. L. D. Faddeev, *Sov. Phys. JETP* **12**, 1014 (1961); *Sov. Phys. Doklady* **6**, 384 (1962).
26. G. Breit, *Phys. Rev.* **120**, 287 (1960).
27. S. N. Gupta, *Phys. Rev.* **117**, 1146 (1960).
28. A. Kallio and K. Kolltveit, *Nucl. Phys.* **53**, 87 (1964).

And Don't Forget the Black Holes

Hans A. Bethe, Gerald E. Brown and Chang-Hwan Lee

Commentary by G. E. Brown

In 2005 we wrote this article to draw attention to the Bethe and Brown article, "Evolution of Binary Compact Objects That Merge."[1] In that paper we had shown that the standard model for evolving binary neutron stars always led to a low-mass black-hole, neutron-star binary once one allowed for (hypercritical) accretion during the pulsar, red-giant common envelope evolution after the pulsar was born while companion giant was evolving. A special way in which to evolve binary neutron stars had already been developed by Brown,[2] which required the two giant progenitors to be sufficiently close in mass ($\lesssim 4\%$) so that they burned He at the same time. The common envelope involving a neutron star was in that way avoided. This special way was estimated to produce only $\sim 10\%$ as many double neutron star binaries as the more common low-mass black-hole, neutron-star binaries.

We submitted this article to *Nature*: it was returned to us the next day, without refereeing, with the statement that it was not of sufficient general interest to warrant publication. Then in the 6th October, 2005 issue of *Nature* four letters, each with a number of authors, and a commentary by Luigi Piro on short hard gamma-ray bursts appeared, just the mergings we had predicted.

Ironically I wrote about this on p. 64 of the October 2005 issue of *Physics Today*, reprinted in this volume, saying that I thought our paper[1] in which we worked things out quantitatively was "the best thing Hans and I did together." I also said that it and the papers that followed are mostly about the future. I didn't realize that some of the future would come out in the same month after the Bethe memorial volume of *Physics Today*.

One nice development since the discovery of the short-hard gamma-ray bursts, which we interpret as coming almost completely from our low-mass black-hole, neutron-star mergings, is in the case made in the paper by Nakar et al.[3] that the merging binaries were formed at about half the Hubble time, $\tau \approx 6$ Gyr ago. At that time the rate of star formation was an order of magnitude greater than at present in our Galaxy, so that the total number of short-hard gamma-ray bursts should be increased by this factor. In other words, the predicted $\sim 10^{-4}$ per galaxy per year merging rate of Bethe and Brown (see below) is increased to $\sim 10^{-3}$ per galaxy per year. In the Nakar et al. units, this is[a] 10^4 mergers Gpc^{-3} yr^{-1}, whereas their "best guess" at the local rate is $\sim 2 \times 10^3$ Gpc^{-3} yr^{-1} with assumption of no beaming, and possibly ~ 10–50 times greater with beaming.

The article by Ehud Nakar on the short-hard gamma-ray bursts which will come out in the Bethe Centennial Volume of *Physics Reports* should give a more complete summary of the short-hard gamma-ray burst situation.

Pinsonneault and Stanek[4] wrote an extremely interesting criticism of our work, pointing out that binaries like to be twins, with the net result that our maximum ratio of low-mass black-hole, neutron-star binaries is only half of what we predicted. By "twins" is meant that the number of binaries with mass ratio q near unity is greatly increased over what one would estimate for uncorrelated Salpeter mass functions.

To our knowledge this paper was the first one to accept the Brown[2] and Bethe and Brown[1] way of making neutron star binaries through double helium burning.

We would like to make two remarks here. Firstly, we are overjoyed that someone has finally accepted the Brown[2] double He star scenario for evolving double neutron stars, which Bethe and Brown[1] made quantitative. Secondly, as far as LIGO and short-hard gamma-ray bursts are concerned, the greater number of neutron star binaries, essentially replaces the lowered number of low-mass black-hole, neutron-star binaries.

In other words, the evolution of binaries of compact objects proceeds along parallel channels. If one channel is made easier, as the evolution of binary neutron stars through the enhancement by twins, then there is less evolution in the other channel, that of low-mass black-hole, neutron-star channel.

---

[a]Assuming 10^5 galaxies within 200 Mpc.

References

1. H. A. Bethe and G. E. Brown, "Evolution of binary compact objects that merge," *Astrophys. J.* **506**, 780–789 (1998).
2. G. E. Brown, "Neutron star accretion and binary pulsar formation," *Astrophys. J.* **440**, 270–279 (1995).
3. E. Nakar, A. Gal-Yam and D. B. Fox, "The local rate and the progenitor lifetime of short-hard gamma-ray bursts: synthesis and predictions for LIGO," astro-ph/0511254, 2005.
4. M. H. Pinsonneault and K. Z. Stanek, "Binaries like to be twins: implications for doubly degenerate binaries, the type Ia supernova rate, and other interacting binaries," *Astrophys. J.* **639**, L67–L70 (2006).

And Don't Forget the Black Holes

Hans A. Bethe, Gerald E. Brown and Chang-Hwan Lee

The discovery of the highly relativistic neutron star (NS) binary (in which both NS's are pulsars) not only increases the estimated merging rate for the two NS's by a large factor, but also adds the missing link in the double helium star model of binary NS evolution. This model gives ~ 20 times more gravitational merging of low-mass black-hole (LMBH), NS binaries than binary NS's, whatever the rate for the latter is.

The recent discovery of Burgay et al.[1] of the double pulsar PSR J0737−3039A and PSR J0737−3039B[2] is very interesting for many reasons. One that is not so obvious is that it involves just about the lowest possible mass main sequence giants as progenitors, which have not been encountered in the evolution of the other binary NS's, even though there are 20 to 30 times more of them, as we show below.

The observational evidence is strong that NS binaries evolve from double helium stars, avoiding the common envelope evolution of the standard scenario.[3] (A helium star results in a giant when the hydrogen envelope is lifted off — in binary evolution by being transferred to the less massive giant in the binary.) In the standard scenario of binary NS formation after the more massive giant transfers its hydrogen envelope to the companion giant, the remaining helium star burns and then explodes into a NS. In about half the cases the binary is not disrupted in the explosion. The NS waits until the remaining giant evolves (and expands) in red giant following its main sequence hydrogen burning. Once the envelope is close enough to the NS, the latter couples to it hydrodynamically through gravity. In the system in which the NS is at rest, it sees the envelope matter coming at it. Some of it is accreted onto the NS, although most flies by in the wake, being heated in the process, and is lost into space. The energy to expel the matter comes from the drop in potential energy as the orbit of the NS tightens. Formulas for the tightening and the amount of mass accreted by the NS were given by Bethe and Brown.[4]

Chevalier[5] first estimated that in the common envelope evolution the NS would accrete sufficient matter to evolve into a black hole (BH). This was made quantitative by Bethe and Brown[4] who calculated that in a typical case, the NS would accrete $\sim 1 M_\odot$, taking it into an $\sim 2.4 M_\odot$ low mass black hole (LMBH), similar to the BH we believe resulted from SN1987A.

Thus, if the NS had to go through common envelope evolution in a hydrogen envelope of $\gtrsim 10 M_\odot$ from the giant, it would accrete sufficient matter to go into a LMBH. Therefore, when the giant evolved into a helium star, which later exploded into a NS, a LMBH–NS binary would result provided the system was not broken up in the explosion. (About 50% of the time the system survives the explosion.) The above scenario was estimated[4] to take place 10 times more frequently than binary NS formation which required the two stars to burn helium at the same time, and, because of the greater mass of the BH, the mergings of binaries with LMBH are twice as likely to be seen as those with only NS's. This is the origin of the factor 20 enhancement of gravitational mergers to be observed at LIGO, over the number from binary NS's alone.

We review the known NS binaries and show that they are consistent with the two NS's in a given binary having very nearly the same mass, as would follow from the double helium star scenario.

We list in Table 1 the 5 observed NS binaries with measured masses. The very nearly equal masses of pulsar and companion in 1534+12 and 2127+11C is remarkable. We show below that 1913+16 comes from a region of giant progenitors in which the masses could easily be as different as they are. The uncertainties in 1518+49 are great enough that the masses could

Table 1. Five observed neutron star binaries with measured masses.

Object	Mass ($M_\odot$)	Object	Mass ($M_\odot$)	Refs.
J1518+4904	$1.56^{+0.13}_{-0.44}$	J1518+4904 companion	$1.05^{+0.45}_{-0.11}$	6, 7
B1534+12	$1.3332^{+0.0010}_{-0.0010}$	B1534+12 companion	$1.3452^{+0.0010}_{-0.0010}$	8
B1913+16	$1.4408^{+0.0003}_{-0.0003}$	B1913+16 companion	$1.3873^{+0.0003}_{-0.0003}$	9
B2127+11C	$1.349^{+0.040}_{-0.040}$	B2127+11C companion	$1.363^{+0.040}_{-0.040}$	10
J0737−3039A	$1.337^{+0.005}_{-0.005}$	J0737−3039B	$1.250^{+0.005}_{-0.005}$	2

well be equal. The main point in our letter is to show that the masses in the double pulsar J0737−3039A and J0737−3039B were probably be very nearly the same before a common envelope evolution in which the first formed NS J0737−3039A accreted matter from the evolving (expanding) helium star progenitor of J0737−3039B in the scenario of Dewi and van den Heuvel.[11]

Since the helium burning in the giant is an order of magnitude shorter than the hydrogen burning in time, for two giants to burn helium at the same time they have to be very nearly equal in mass, within 4%. Thus the binary neutron star scenario is highly selective. If they do burn helium at the same time, they can go through common envelope evolution at this time. The hydrogen envelope of the slightly more massive giant expands in red giant and is transferred to the less massive giant, which in turn evolves into red giant. The time for these evolutions is so short that the helium stars are unable to accept the hydrogen which is lost into space[12] so that the final neutron stars are also close in mass. Each of the helium stars will explode, going into a NS. In about half of each explosion the binary is disrupted, but in about 1/4 of the cases it is preserved and a double neutron star results.

Bethe and Pizzochero[13] used for SN 1987A a schematic but realistic treatment of the radiative transfer problem which allowed them to follow the position in mass of the photosphere as a function of time. They showed that the observations determine uniquely the kinetic energy of the envelope once the mass is known. From the envelope masses considered, the range of energies was 1 to 1.4×10^{51} ergs. Using the fact that the pressure following the supernova shock is radiation dominated, Bethe and Brown[14] showed from the known value of $0.075 M_\odot$ of ^{56}Ni production, an upper limit on the gravitational mass of $1.56 M_\odot$ could be obtained. This is just the calculated Fe core mass for the known $18 M_\odot$ progenitor of 1987A.

Calculations of Woosley are shown in Table 3 of Brown et al.[15] where the amounts of fallback material from a distance of 3500 km and 4500 km are given. The fallback simply cancels the gravitational binding energy from the Fe core evolving into an NS. Since the density of matter is thin at the estimated 4000 km bifuraction point, this cancellation is relatively insensitive to the precise point of bifurcation. Thus, we can use the Fe core mass as the mass of the NS.

In Fig. 1, we show both recent and older calculations of Fe core masses. The filled circles and crosses correspond to core masses at the time of iron core implosion for a finely spaced grid of stellar masses.[16] The circles were

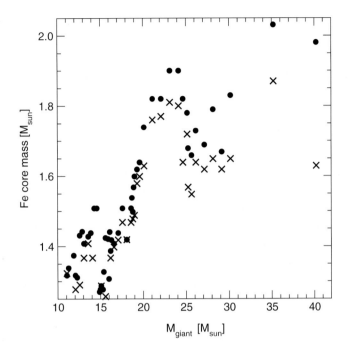

Fig. 1 Fe core masses for a grid of stellar masses. See text for explanation.

calculated with the Woosley and Weaver code,[17] whereas the crosses employ the vastly improved rates for electron capture and beta decay.[18]

A rapid increase in Fe core masses occurs at $M_{\rm giant} \sim 18 M_\odot$, just the mass of the progenitor of 1987A which we believe went into a LMBH. The Burrows and Woosley[19] evolution of 1913+16 took place in terms of $M_{\rm giant} \sim 20 M_\odot$. (In this evolution the Fe core mass will be slightly less than given in Fig. 1 because the mass transfer in binary evolution removes the hydrogen envelope, leaving a "naked" helium star. The wind loss from naked helium stars is great.) Now one can see that the difference in masses between, 19 and $20 M_\odot$ giants is greater than $0.1 M_\odot$, so that the $0.05 M_\odot$ difference in masses between the pulsar and its companion in 1913+16 can easily be furnished out of the calculated Fe core masses.

There is some tendency for the Fe core masses to decrease from $M_{\rm giant}$ of 15 to $16 M_\odot$, where we would estimate 1534+12 and its companion to come from. (The companion is slightly more massive than the pulsar.)

The pulsar and companion in the double pulsar should come from $M_{\rm giant} \sim 10 M_\odot$, at the $1.25 M_\odot$ of the present J0737−3039A. Following Dewi and van den Heuvel,[11] the first pulsar formed is increased in mass with mass transfer from the companion when the latter is a helium star. Low-mass helium stars, 2.3 to $3.3 M_\odot$ have to burn hotter than higher mass ones because of the greater energy loss from their surfaces. In reaction to this, they expand in a helium red giant phase. The pulsar now goes through a common envelope stage with the envelope of the helium star once the latter expands far enough to make contact.

As noted, the NS accretes some of the matter and most is expelled in the wake. Thus, from the coefficient of dynamical friction $c_d = 2\ln(b_{\max}/b_{\min}) \simeq 6$,[20] it is found[4] that

$$\frac{M_{\text{pulsar,f}}}{M_{\text{pulsar,i}}} = \left(\frac{M_{\text{He,f}}/a_f}{M_{\text{He,i}}/a_i}\right)^{1/5}. \qquad (1)$$

We take $M_{\text{He,i}} = 2.5 M_\odot$ and $M_{\text{He,f}} = 1.5 M_\odot$. Now the orbit in the double pulsar is decreased a factor of $\sim 3.7^{2/3}$ from that of the binary NS. (The average of periods of 1913+16 and 1534+12 is 0.37 days, whereas that of the double pulsar is 0.1 day.) The 2/3 exponent is the Kepler relation between orbital period and radius, and we then find

$$\frac{M_{\text{pulsar,f}}}{M_{\text{pulsar,i}}} = 1.075 \qquad (2)$$

which gives the correct $1.34 M_\odot$ for J0737−3039B given $M_{\text{pulsar,i}} = 1.25 M_\odot$, the same as the unrecycled neutron star. The $1.5 M_\odot$ He star will burn into the unrecycled pulsar, with mass decreased $\sim 15\%$ by the general relativity binding energy.

Therefore, in the double pulsar, the masses of the two giant progenitors must be very close, since the two NS's would have had nearly equal masses (as in 1534+12), except for the mass transfer during the helium red giant stage.

The Salpeter mass distribution of giants gives the relative number proportional to $(M_{\text{giant}})^{-2.35}$, so for two equal mass giants, the distribution goes as $(M_{\text{giant}})^{-4.7}$. Thus, the probability of two $10 M_\odot$ giants is 26 times greater than that of two $20 M_\odot$ giants, as would be progenitors of 1913+16. Burgay et al.[1] estimated the ratio N_{0737}/N_{1913} to peak at ~ 6, whereas Dewi and van den Heuvel[11] estimate that there are many of the PSR J0739-like systems not seen because of their weak signals compared with those from 1913+16, possibly giving a factor ~ 30 in mergings. Our estimate based on the large number of low-mass giant progenitors confirms this.

With $\gtrsim 10 M_\odot$ hydrogen envelopes and a proportional tightening of the orbit in the common envelope evolution in the standard scenario, the Bethe and Brown result of accretion of $\sim 1 M_\odot$ onto the first NS seems reasonable.

After only part of this is accreted, the neutron star, like in 1987A, goes into a black hole.

Why haven't we seen any LMBH–NS binaries? Van den Heuvel[21,22] has pointed out that NS's form with strong magnetic fields 10^{12} to 5×10^{12} gauss, and spin down in a time

$$\tau_{\rm sd} \sim 5 \times 10^6 \text{ years} \qquad (3)$$

and then disappear into the graveyard of NS's. (The pulsation mechanism requires a minimum voltage from the polar cap, which can be obtained from $B_{12}/P^2 \sim 0.2$ with $B_{12} = B/10^{12}$G and P in seconds.[22]) The relativistic binary PSR 1913+16 has a weaker field $B \simeq 2.5 \times 10^{10}$ gauss and therefore emits less energy in magnetic dipole radiation. Van den Heuvel estimates its spin-down time as 10^8 yrs. There is thus a premium in observational time for lower magnetic fields, in that the pulsars can be seen for longer times.

Taam and van den Heuvel[23] found empirically that the magnetic field of a pulsar dropped roughly linearly with accreted mass. This accretion can take place from the companion in any stage of the evolution. A pulsar that has undergone accretion is said to have been "recycled."

Now as mentioned, in 1913+16 the pulsar magnetic field is $\sim 2.5 \times 10^{10}$ gauss and in 1534+12 it is $\sim 10^{10}$ gauss. In J0737−3039A it is only 6.3×10^9 gauss. These "recycled pulsars" will be observable for $\gtrsim 100$ times longer than a "fresh" (unrecycled) pulsar.

The same holds for LMBH–NS binaries. The NS is certainly not recycled, so there is about 1% chance of seeing one as the recycled pulsar in a binary NS. But we propose 10 times more of the LMBH's than binary NS's, of which we observe 5. Thus the total probability of seeing the LMBH binary should be about 50%. However, there may be additional reasons that the LMBH–NS binary is not observed. NS's in binaries in which the nature of the companion star is unknown are observed. We may have to wait for LIGO which will be able to measure "chirp" masses quite accurately. The chirp mass of a NS binary should concentrate near $1.2 M_\odot$, whereas the LMBH–NS systems should have a chirp mass of $1.6 M_\odot$, and there should be ~ 20 times more of the latter.

In order to estimate the number of mergings per year for the LIGO we begin from the estimate of Kim et al.[24] and multiply by a factor of 26 to account for the increase in this estimate by the large number of low-mass giant progenitors, neutron stars from two of these having been observed in the double pulsar. Additionally we multiply by the factor of 20 to include NS–LMBH binary mergings. This brings our estimate up to 0.5 to 3.6 mergings per year for the initial LIGO, and about 6000 times greater for the advanced LIGO.

Acknowledgments

G.E.B. was supported in part by the US Department of Energy under Grant No. DE-FG02-88ER40388. C.H.L. was supported by the Korea Research Foundation Grant funded by the Korea Government (MOEHRD)(KRF-2005-070-C00034).

References

1. M. Burgay et al., "An increased estimate of the merger rate of double neutron stars from observations of a highly relativistic system," Nature **426**, 531–533 (2003).
2. A. G. Lyne et al., "A double-pulsar system: a rare laboratory for relativistic gravity and plasma physics," Science **303**, 1153–1157 (2004).
3. E. P. J. Van den Heuvel and J. van Paradijs, "X-ray binaries," Scientific American November, 38–45 (1993).
4. H. A. Bethe and G. E. Brown, "Evolution of binary compact objects that merge," Astrophys. J. **506**, 780–789 (1998).
5. R. A. Chevalier, "Neutron star accretion in a stellar envelope," Astrophys. J. **411**, L33–L36 (1993).
6. S. E. Thorsett and D. Chakrabarty, Astrophys. J. **512**, 288–299 (1999).
7. D. J. Nice, R. W. Sayer and J. H. Taylor, Astrophys. J. **466**, L87–L90 (1996).
8. I. H. Stairs, S. E. Thorsett, J. H. Taylor and A. Wolszczan, Astrophys. J. **581**, 501–508 (2002).
9. J. M. Weisberg and J. H. Taylor, in Radio Pulsars, eds. M. Bailes, D. J. Nice and S. Thorsett, Astronomical Society of the Pacific, San Francisco, 2003, pp. 93–98.
10. W. T. S. Deich and S. R. Kulkarni, in Compact Stars in Binaries: IAU Symposium 165, eds. J. van Paradijs, E. P. J. van den Heuvel and E. Kuulkers, Kluwer, Dordrecht, 1996, pp. 279–285.
11. J. D. M. Dewi and E. P. J. van den Heuvel, "The formation of the double neutron star pulsar J0737 − 3039," Mon. Not. R. Astron. Soc. **349**, 169–172 (2004).
12. H. Braun and N. Langer, "Effects of accretion onto massive main sequence stars," Astron. Astrophys. **297**, 483–493 (1995).
13. H. A. Bethe and P. Pizzochero, "Mass-energy relation for SN 1987A from observations," Astrophys. J. **350**, L33–L35 (1990).

14. H. A. Bethe and G. E. Brown, "Observational constraints on the maximum neutron star mass," *Astrophys. J.* **445**, L129–L132 (1995).
15. G. E. Brown, J. F. Weingartner and R. A. M. J. Wijers, "On the formation of low-mass black holes in massive binary stars," *Astrophys. J.* **463**, 297–304 (1996).
16. A. Heger, S. E. Woosley, G. Martinez-Pinedo and K. Langanke, "Presupernova evolution with improved rates for weak interactions," *Astrophys. J.* **560**, 307–325 (2001).
17. S. E. Woosley and T. A. Weaver, "The evolution and explosion of massive stars. II explosive hydrodynamics and nucleosynthesis," *Astrophys. J.* **101**, 181–235 (1995).
18. K. Langanke and G. Martinez-Pinedo, "Shell-model calculations of steller weak interaction rates: II. Weak rates for nuclei in the mass range A = 45−65 in Supernovae Environments," *Nucl. Phys.* **A673**, 481–508 (2000).
19. A. Burrows and S. E. Woosley, "New constraints on the progenitor of the binary pulsar — PST 1913+16," *Astrophys. J.* **308**, 680–684 (1986).
20. E. Shima, T. Matsuda, H. Takeda and K. Sawada, "Hydrodynamic calculations of axisymmetric accretion flow," *Mon. Not. R. Astron. Soc.* **217**, 367–386 (1985).
21. E. P. J. Van den Heuvel, "Interacting binaries: topics in close binary evolution," in *Lecture Notes of the 22nd Advanced Course of the Swiss Society for Astronomy and Astrophysics (SSAA)*, eds. H. Nussbaumer and A. Orr, Springer, Berlin, 1994, pp. 263–474.
22. D. Bhattacharya and E. P. J. Van den Heuvel, "Formation and evolution of binary and millisecond radio pulsars," *Phys. Reports* **203**, 1–124 (1991).
23. R. E. Taam and E. P. J. van den Heuvel, "Magnetic field decay and the origin of neutron star binaries," *Astrophys. J.* **565**, 235–245 (1986).
24. C. Kim, V. Kalogera and D. R. Lorimer, "The probability distribution of binary pulsar coalescence rates. I. Double neutron star systems in the galactic field," *Astrophys. J.* **584**, 985–995 (2003).

Shaping Public Policy*
Sidney Drell

I am of the generation that entered into serious study of physics just after World War II. I soon realized that some of the great names of modern physics, from whose theories and experiments I was learning the most exciting new stuff, were prominently in the news. They were speaking, debating, testifying in Congress on issues of grave policy importance concerning the impact on the human condition of the new technologies spawned by the latest scientific progress, and discussing what our government should be doing about it. Nuclear weapons were of special concern in this regard with their enormous destructive potential that posed a threat to civilization as we know it, in the event of a nuclear war.

As students and young aspiring researchers, we had many discussions on civilian vs. military control of nuclear weapons; whether or not to build the super, or hydrogen bomb; was it appropriate for the government to require security clearances for its fellowships for us to get through graduate school. These questions were ever present, and it didn't take very long for us to recognize that, amidst many grandiose, often hyperbolic statements about the challenges we faced, one voice came through straightforward, clear, firm, and on target. It was that of Hans Bethe, and it carried an authority and weight just like his writings and lectures in physics with which we were so familiar: the Bethe/Heitler formula; the Bethe/Bloch energy loss formula; the Bethe Bible on nuclear physics; his theory explaining the production of energy in the burning sun that sustains life on earth; the Lamb shift. Just as Bethe's physics went directly, clearly and logically to the point, so did

---

*Presented at the Hans Bethe Symposium, Los Alamos National Laboratory, 19 August 2005.

his analyses, based on his vast scientific and technical knowledge, of new problems presented by nuclear weapons and missiles.

We listened carefully when Hans Bethe testified or wrote about building the H-bomb; about what could and should be done to limit the build up of nuclear weapons as well as their spread to other nations; when he explained the serious technical limitations of ballistic missile defenses as well as the relative ease of countering their postulated effectiveness; and when he spoke on stopping test explosions of nuclear weapons, particularly in the atmosphere.

It wasn't until 1960 that I first became actively involved in actually working on technical aspects of those issues. That came with the formation of JASON, a group of then young, and at least to some, promising scientists. We were academics invited to work, at summer studies and in our spare time, on important technical issues of national security. This gave me my first opportunity to meet Hans, who served as a senior adviser in those early days of JASON. That was the beginning of a great friendship, in Humphrey Bogart's unforgettable closing words in *Casablanca*; indeed a cherished friendship that further deepened my great admiration and respect for Hans. Personal friendships are hard to describe. But I have one quantitative measure of the extent of this one. I can think of no one, outside of my immediate family members, with whom, over the past 40 years, I have had more meals than with Hans, whether working in Washington, Livermore, or visiting Ithaca frequently during the years my daughter was there, or wherever.

Hans had an important, and at times decisive, impact on just about every major issue involving nuclear weapons policy that called for scientific and technical knowledge and judgment. He gave unstintingly of his enormous scientific talents to help the United States government make wise policy choices when it came to building a safe and reliable nuclear deterrent, to negotiating and verifying arms control treaties, or to understanding technical limits on complex systems. Hans approached technical problems with an open mind. His integrity, together with his extraordinary scientific knowledge, gave great credibility to his input in policy decisions, and made his advice to the government and his public testimony uniquely valuable. Unfortunately in this role one is not always listened to. The batting average in activities of this type, measured by the percentage of success in moving policy decisions in the desired directions, is not high, nowhere near as high as one would hope, and as it should be. This never discouraged Hans. He would just work harder the next time.

Hans also had the intellectual integrity and courage to stand up and speak out in criticizing the government and officials in defense of colleagues that he felt were being attacked unjustly. This was evident in his ringing defense of Robert Oppenheimer whose loyalty was impugned during the communist witch-hunting days in the early 1950's. When asked during Oppie's loyalty hearings "to express an opinion about Dr. Oppeheimer's loyalty to the United States, about his character, about his discretion in regard to matters of security," Hans pulled no punches in his response:

> I am certainly happy to do this. I have absolute faith in Dr. Oppenheimer's loyalty. I have always found that he had the best interests of the United States at heart. I have always found that if he differed from other people in his judgment, that it was because of a deeper thinking about the possible consequences of our action than the other people had. I believe that it is an expression of loyalty — of particular loyalty — if a person tries to go beyond the obvious and tries to make available his deeper insight, even in making unpopular suggestions, even in making suggestions which are not the obvious ones to make and are not those which a normal intellect might be led to make.
>
> I have absolutely no question that he has served this country very long and very well. I think everybody agrees that his service in Los Alamos was one of the greatest services that were given to this country. I believe he has served equally well the GAC (the General Advisory Committee of the AEC) in reestablishing the strength of our atomic weapons program in 1947. I have faith in him quite generally.

Hans was no less strong, persistent, and courageous in defending less well-known colleagues he knew and trusted when they were attacked in Congressional hearings, viciously and often falsely, for past political associations.

In presenting his advice and stating his views, Hans conveyed the strength of a realist with deep moral feelings and ideals. In his case it was the special strength of one who had challenged his ideals against troubling realities and inescapable ambiguities that we often must confront in today's world. Consider Hans' words in struggling with the challenges, dangers, and opportunities of moving ahead to build thermonuclear weapons — the so-called hydrogen bombs. In 1954 he wrote an article entitled "Comments on the History of the H-Bomb." Originally classified, it wasn't released until 1982 when it was published in the magazine *Los Alamos Science*. In it he quoted a conclusion from the report of the GAC on proceeding to build the H-bomb:

> We all hope that by one means or another, the development of these weapons can be avoided. We are all reluctant to see the United States take the initiative in precipitating this development. We are all agreed that it would be wrong at the present moment to commit ourselves to all-out efforts toward its development.

Bethe then commented that the GAC report

> might well be considered as a prayer for some solution to the dilemma, not as an answer. Scientists are not especially qualified to find a solution in the domain of statecraft. All they could do was to point out that here was a very major decision and it was worth every effort to avoid an irrevocable, and perhaps fatal, step.

And he concluded:

> In summary I still believe that the development of the H-bomb is a calamity. I still believe that it was necessary to make a pause before the decision and to consider this irrevocable step most carefully. I still believe that the possibility of an agreement with Russia not to develop the bomb should have been explored. But once the decision was made to go ahead with the program, and once there was a sound technical program, I cooperated with it to the best of my ability. I did and still do this because it seems to me that once one is engaged in a race, one clearly must endeavor to win it. But one can try to forestall the race itself.

Later, in 1968 in an interview for an article in *The New York Times Magazine*, Hans reflected on his decision to work on the H-bomb as follows[a]:

> Just a few months before, the Korean war had broken out, and for the first time I saw direct confrontation with the Communists. It was too disturbing. The cold war looked as if it were about to get hot. I knew then I had to reverse my earlier position. If I didn't work on the bomb somebody else would — and I had the thought if I were around Los Alamos I might still be a force for disarmament. So I agreed to join in developing the H-bomb.
>
> It seemed quite logical. But sometimes I wish I were more consistent an idealist.

---

[a]S. S. Schweber, *In the Shadow of the Bomb: Bethe, Oppenheimer, and the Moral Responsibility of the Scientist*, Princeton University Press, 2000, p. 166.

One can only admire an individual who forthrightly confronts the dilemmas, the ambiguities, the occasional inconsistencies that must be resolved in making difficult decisions — and, after fully acknowledging that they exist, makes the hard choices and acts on them!

Throughout four decades of the Cold War, Hans advocated and worked to create effective tools for verifying and validating negotiated agreements to slow the nuclear arms competition between the United States and the former Soviet Union. As one of the original members of the President's Science Advisory Committee (PSAC), established by President Eisenhower in 1957, Hans proposed a study of a ban on nuclear weapons tests. In typical Bethe fashion the initial work was intended to focus on specific and very important technical questions: What effect would such a ban, or a temporary moratorium, have on the US and Soviet programs? And, would it be possible to monitor compliance by being able to detect explosive tests in violation of such a ban? The original goal was a comprehensive ban on explosive testing in all environments. If it could be established on technical grounds that compliance with such limitations could be verified, the hope was to engage with the Soviets and British in exploring the prospects for such an agreement.

James Killian, the President's first Science Advisor and Chairman of PSAC, appointed an interagency panel, chaired by Hans, to make technical assessments on the ability of the US to detect Soviet nuclear tests under various conditions, and to assess how such constraints on testing would affect the weapons laboratories and their work, as well as the then-existing Soviet and US nuclear arsenals. The panel did their work well and responsibly, describing a practical detection system that would be able to identify nuclear explosions except for very low-yield underground detonations. However, they relegated to a more appropriate and broader forum the judgment of whether a test ban would be in the net military interests of the United States or of the Soviet Union, recognizing in particular that such restrictions would delay if not prevent additional testing of "clean" small and relatively inexpensive nuclear weapons. One hears echoes of that concern in some of today's debates on building new weapons.

This work of the Bethe panel led to a conference of technical experts from the United States, Great Britain, and the Soviet Union meeting in Geneva during the summer of 1958. There were serious — excessively serious — concerns by the Soviets about the motives of the Western negotiators and the necessity of preserving the secrecy surrounding their own nuclear program. There were also intense debates within the United States as committed opponents of any such ban, led by Lewis Strauss, Ernest Lawrence, and Edward Teller, insisted that nothing less than an unrestricted, on-site, challenge inspection system could guarantee that the Soviets were not cheating.

The resolution of the matter of a test ban came five years later, in 1963, when the United States, Great Britain, and the Soviet Union signed a treaty banning nuclear weapons tests in the atmosphere, in outer space, and in the oceans, but allowing underground explosive tests that spread no dangerous radioactive fallout to continue. Hans was disappointed that the negotiators failed to reach an agreement that would permit underground tests below a low-yield threshold, above which they could be detected by long-range seismic signals. But he also recognized the importance of his efforts in ending the era of extensive atmospheric testing, and in having issues of arms reduction discussed and taken seriously at the highest level of the US government. In his book, Schweber comments (on page 177) that Hans considered this to be his most satisfying and important contribution as a scientist in the public arena.

Killian summarized Bethe's contributions as follows in his 1977 memoir *Sputnik, Scientists, and Eisenhower*[b]:

> Bethe, a man of tremendous intellect, an idealist with a whiplash mind ... was to become one of the heroes of the long campaign that led to the test ban of 1963 ... Bethe possesses a grave nobility of character that has commanded the respect and affection of all who have worked with him.
>
> With these qualities and his deep knowledge of nuclear matters, Bethe gained the confidence of the interagency committee and directed its work with skill. ... the members worked well together and some even changed their views during the study.

In 1966, on the occasion of Bethe's 60th birthday, President Lyndon Johnson summarized why Hans had become a national treasure with these words in a letter he sent to him:

> You are not only an outstanding scientist, you are also a devoted public servant.
>
> The nation has asked for your help many times and you have responded selflessly. You have made profound contributions in the fields of atomic energy, arms control and military technology. And

---

[b]J.R. Killian, *Sputnik, Scientists, and Eisenhower*, MIT Press, Cambridge, Mass., 1977, pp. 154-155.

you have been an important source of the immense contribution which science and the university community are making to society as a whole.

Our country is deeply indebted to you.

The work of the Bethe panel starting in 1958 that culminated in the Limited Test Ban Treaty of 1963 is but one example of the enormously valuable advice that a strong scientific advisory mechanism in the White House can provide for the president and his staff. President Eisenhower recognized this in 1957 when he established PSAC as his resource for direct, in-depth analyses and advice as to what to expect from science and technology, both current and in prospect, in establishing realistic national policy goals. Two things set PSAC apart from the existing governmental line organizations and cabinet departments with operational responsibilities, as well as from nongovernmental organizations engaged in policy research. First of all, its members had White House backing and the requisite security clearances to gain access to all the relevant information for their studies on highly classified national security issues. Secondly, the individual members, selected on grounds of demonstrated achievements in science and engineering, were independent and presumably, therefore, immune from having their judgments affected by operational and institutional responsibilities. Therein lies their unique value.

Highly respected, leading American scientists like Ed Purcell, Hans Bethe, Pief Panofsky, Din Land, and George Kistiakovsky made contributions of enormous value to US security as we grappled with new threats posed by nuclear weapons and intercontinental range missiles. They were also instrumental in pushing forward technology for gaining strategic reconnaissance from space. Starting with U2 overflights of the Soviet Union in 1956, followed by the enormous technical leap into space in 1960 with CORONA, the first photo reconnaissance satellite, we pierced the Soviet's iron curtain, making it possible to accurately assess new threats that might be developing as well as to dispel some that we incorrectly assumed to exist. This was especially critical in an era when the danger of surprise nuclear attack was then only 30 minutes away, which is the ICBM flight time, and we were no longer shielded by large oceans. Of equal importance, with the ability to monitor the growth of the Soviet strategic nuclear threat in numbers of large nuclear armed missiles, bombers, and submarines, we could initiate negotiating efforts to limit the arms race — and eventually reverse it — based on the essential ability to verify compliance with treaty provisions. It is too bad that there no longer exists a mechanism comparable to the original PSAC that is providing the White House today with the kind of high level independent scientific advice that proved so valuable back in 1957.

But to get back to the important contributions of Bethe, they continued unabated for decades following the Limited Test Ban Treaty of 1963. Throughout the 1970's and 80's other issues replaced a nuclear test ban as the main focus of public attention. The most important of these were the ABM Treaty of 1972 and the NonProliferation Treaty that entered into force in 1970. Bethe was a particularly active participant in the debates on ABM systems. He recognized their inherent technical limitations relative to countermeasures readily available to the offensive missile forces. Moreover, in view of the enormous destructive potential of nuclear weapons, the level of perfection that was required to provide protection for civilian populations was impossible to achieve. In 1968, together with Richard Garwin, he published the first comprehensive and unclassified analysis on the fundamental technical limitations of ballistic missile defense. That article in the *Scientific American* played a major role in informing and framing the public debate.

The ABM debate was waged with extraordinarily fierce passion and intensity because it challenged a deep human instinct. We had to recognize that, against all history of the pre-nuclear era, we now had to accept the conclusion that protecting people, our families and friends, and cities in the nuclear age was no longer possible. The technical realities dictated this conclusion, and thus scientists and engineers were critical participants in the public policy debates. Whatever we may prefer as the goal of our policy, we cannot deny or evade the laws of nature!

During those years efforts to negotiate further testing limitations on nuclear weapons continued. In 1974 a small step of progress was made in the agreement to limit underground nuclear explosions to yields no larger than 150 kilotons, which is 10 times larger than the Hiroshima bomb. But the testing issue did return to center stage in 1992 for both technical and political reasons. Technically it was broadly recognized that the existing US nuclear arsenal was reliable, met established and very strict safety criteria, and was effective, and hence there was no strong reason to continue underground nuclear explosive tests. This in turn generated strong congressional pressure to cease underground tests, in response to which the first President Bush initiated a one-year moratorium on such tests in 1992. In addition, political pressure for an end to testing was growing around the world. It became increasingly vocal at the regular 5-year reviews of the NonProliferation Treaty at the United Nations, and particularly so in 1995 at its final scheduled 5-year review.

At that occasion, many nations criticized the Treaty's discriminatory features, and, as a *quid pro quo* for their continuing to renounce nuclear weapons, called on the nuclear powers to make serious and timely progress in reducing their excessively large arsenals and their reliance on nuclear weapons. They also called on them to continue to work toward a Comprehensive Test Ban Treaty (CTBT) that would formalize the existing moratorium on testing and extend it without a limit of time.

It is in this context that the United States must evaluate the impact of new initiatives to develop a new generation of nuclear warheads, an issue that has been under serious consideration for the past two years in Washington. Will the non-nuclear nations be willing to continue their policy to forego developing them if we, the strongest nation in the world, by far, the only super-power, insist that our security requires that we develop new ones — bunker-busters or robust nuclear earth penetrators (RNEPs), or new low-yield concepts — for new missions to meet our security needs? What will be the fate of the testing moratorium and of the entire nonproliferation regime if the newly established Reliable Replacement Weapons program evolves into a design program for new weapons, as some suggest, rather than focusing on increasing long term confidence in our current arsenal within experimentally established parameters. Would a responsible leader — President, General, or Admiral — seriously consider relying on an untested new design to protect our national security? What flight of imagination is required to place higher confidence in a new design without a test pedigree than in our stockpile with a half-century of more than 1000 tests in its making?

I think we know very well what Hans would say. He supported the CTBT as consistent with maintaining an effective, reliable and safe deterrent, and important to preserving a nonproliferation regime.

Addressing Los Alamos scientists in 1995, on the 50th anniversary of Hiroshima, Hans issued the following statement urging scientists in all countries to refuse to support the design and development of new improved weapons:

> As director of the Theoretical Division of Los Alamos, I participated at the most senior level in the World War II Manhattan Project that produced the first atomic weapons.
>
> Now, at the age of 88, I am one of the few remaining such senior persons alive. Looking back at the half-century since that time, I feel the most intense relief that these weapons have not been used since World War II, mixed with the horror that tens of thousands of such weapons have been built since that time — one

hundred times more than any of us at Los Alamos could ever have imagined.

Today we are rightly in an era of disarmament and dismantlement of nuclear weapons. But in some countries nuclear development still continues. Whether and when the various nations of the world can agree to stop this is uncertain. But individual scientists can still influence this process by withholding their skills.

Accordingly, I call on all scientists in all countries to cease and desist from work creating, developing, improving and manufacturing further nuclear weapons — and, for that matter, other weapons of potential mass destruction such as chemical and biological weapons.

This sentiment closely reflects an earlier comment made by Andrei Sakharov, the father of the Soviet hydrogen bomb and brave fighter for human rights, on his first visit to the United States in December 1988. Recognizing that he and the people who worked with him at the time "were completely convinced that this work was essential, that it was vitally important," and that "the American scientists in their work were guided by the same feelings of this work being vital for the interests of the country," Sakharov said

> While both sides felt that this kind of work was vital to maintain balance, I think that what we were doing at that time was a great tragedy. It was a tragedy that reflected the tragic state of the world that made it necessary, in order to maintain peace, to do such terrible things. We will never know whether it was really true that our work contributed at some period of time toward maintaining peace in the world, but at least at the time we were doing it, we were convinced this was the case. The world has now entered a new era, and I am convinced that a new approach has now become necessary.

In his last interview before his death, Sakharov also called for a permanent halt of nuclear tests.[c]

Two of the great pioneers of the nuclear age, delving again deep into their moral convictions near the end of their lives, urged the world to recognize that in the post-Cold War world we have entered a new era. Factors other than modernizing nuclear weapons for potential new missions have priority

---

[c]S. D. Drell, *Physics Today*, May 2000.

importance. Our most pressing concern now is the danger of the proliferation of nuclear know-how. As President Bush has remarked: "The gravest danger facing the nation lies at the crossroads of radicalism and technology." Our urgent need is to strengthen and preserve a nonproliferation regime. At stake may very well be the Comprehensive Test Ban Treaty signed on September 24, 1996 at the United Nations, and with it the NonProliferation Treaty.

A point of clarification here: Neither Bethe's call for weapons restraint — nor my own concerns expressed earlier about the future of the NPT and the CTBT — should be interpreted as saying it is time for the weapons labs to close up shop, lock the door and go home. Far from it. The labs and National Nuclear Security Administration have created a very challenging scientific and technical program that is allowing us to maintain confidence in our arsenal by pursuing a broad range of activities that are deepening our basic understanding of nuclear explosions. They include critical experiments, from small tabletop ones to subcrits at the Nevada test site; also validating and verifying high fidelity codes for the new supercomputers that support advanced simulation techniques that can be tested against new data in extreme physical conditions relevant for bomb processes. This data can now, or soon will be obtained from advanced new facilities like DARHT, NIF and Z-pinch pulse power devices. What you are doing is important work. It is good science and is consistent with our national strategic goals to maintain a strong US nuclear deterrent as we downsize it, and at the same time, to continue to maintain the support and cooperation of our non-nuclear allies in our efforts to preserve and strengthen a nonproliferation regime. We must have their active collaboration to enforce effective means of preventing the world's most dangerous people from getting their hands on the world's most dangerous weapons.

Recall the poignant words of President Eisenhower who commented in 1960, upon leaving office, that not achieving a nuclear test ban "would have to be classed as the greatest disappointment of any administration — of any decade — of any time and of any party ..." Upon signing the CTBT in 1996, President Clinton heralded it as "the longest sought, hardest fought prize in the history of arms control." Regrettably the United States has so far failed to ratify the Treaty, although all US allies in NATO have, as have Japan and Russia. In all 173 nations have signed and 120 have ratified the CTBT, including 33 of the 44 so-called nuclear-capable states, i.e. those that have built nuclear reactors and must ratify it for the Treaty to enter into force. When that day comes, it will mark the success of an effort which

Hans pioneered, and in which he played an important role for more than four decades.

Hans Bethe was the last of the founding giants of modern quantum and nuclear physics. He was present at its creation and for more than seven decades contributed enormously to deepening our understanding of the physical nature of the earth and the stars. Beyond his major contributions to advances in modern science and to the development of the atom bomb, he became an important leader among scientists who felt, and acted on, the responsibility of our community to help governments and societies understand the potential impact of these achievements on the human condition. As a government adviser at the highest levels and a participant in public forums he strove to ensure that consequences of scientific and technical advances — particularly in nuclear weapons and energy — were utilized toward peaceful and beneficial purposes.

Ten years ago, in 1995, at a celebration of Hans' 60th anniversary as a professor at Cornell University, Viki Weisskopf gave a perfect one-word description of Hans when he characterized him as "dreadnaught," plowing straight ahead with irresistible force to accomplish his scientific, technical, and policy goals. In the words of his long-time collaborator on astrophysical problems, Gerald Brown, Hans "worked like a bulldozer, heading directly for the light at the end of the tunnel." His achievements led Freeman Dyson to recognize him as the "supreme problem solver of the past century."

Those of us who had the privilege to work with Hans always stood in amazement when difficult questions arose, and out of his mind, which his wife, Rose, once likened to me as a "filing cabinet," came a treasure trove, a veritable catalogue, of the necessary numbers, all the important points, whether they be basic science or applied physics. And Hans could give you, on the spot, a back-of-the-envelope answer that you could bet on.

We admired and respected Hans for what he stood for and what he did. He was a giant in his time. We are all going to miss him.

Hans Bethe and the Global Energy Problems

Boris Ioffe

Bethe's viewpoint on the global energy problems is presented. Bethe claimed that nuclear power was a necessity in future. Nuclear energy had to be based on breeder reactors. Bethe considered the non-proliferation of nuclear weapons as the main problem of the long-term future of nuclear energy. The solution of this problem he saw in heavy water moderated thermal breeders, using uranium-233, uranium-238 and thorium as a fuel.

My Contacts with Bethe

I first heard Bethe's name in 1947. I was a student of the 3rd course of the physics faculty at Moscow University. I was not satisfied with the level of education there, especially with the teaching of theoretical physics. The best Russian theoreticians — Landau, Tamm, Leontovich and others had been expelled from the University, because they did not share the official Marxist philosophy. After some doubts about my ability to become a theoretical physicist I decided to attempt to give up to Landau his theoretical minimum. I passed the first entrance exam in mathematics and Landau gave me the program of study for the other subjects required. At that time only the first three and one-half volumes of Landau's by now well-known *Course of Theoretical Physics* had been published: Mechanics, Classical Field Theory, Mechanics of Condensed Matter and the first part of Statistics (Classical Statistics). For all other parts of the Course the student was expected to study mostly by reading original papers, which were in German and English. (It was implicitly assumed that students knew both languages sufficiently well — a very nontrivial case at that time.) The coursework in Quantum

Mechanics contained three papers of Bethe's: the theory of atomic levels in the fields of crystals[1]; the theory of collisions of fast electrons with atoms[2]; and the theory of the deuteron.[3] It was particularly difficult to study the first two papers — about 75 pages each in German, a language, which I had learned as a young boy and had almost forgotten. To make things even more difficult, the papers contained an enormous amount of complicated calculations. I redid all of them! From that time onwards I have had a great respect for Hans Bethe.

After I had passed all of Landau's examinations, I became a member of the Landau seminar. I met Landau very often, as a rule 2–3 times a week, and I realized that he, too, had a great respect for Bethe. He always referred to him as Hans Albrecht, not by the last name, as was common. (Pomeranchuk called Bethe the same.) For Landau it was a high estimation, if results of any calculations compared well with Bethe's.

I would like to say something, too, about the indirect connection of our work with one of Bethe's insightful remarks, now almost forgotten. In 1951 the Pomeranchuk group at ITEP, in which I participated, started work on the project of the hydrogen bomb, called "Tube" (Truba). This was the continuation of the work, performed earlier by Zeldovich and Landau's groups. In the US a similar project was developed by Teller and was known as the "Classical Super." In short the idea of the project was the following. A long cylinder was filled with liquid deuterium. At one end of the cylinder the atomic bomb and the intermediate device filled by deuterium and tritium were situated. It was expected that after an atomic bomb explosion the fusion reaction would start in the $D+T$ mixture, resulting in high temperatures and that the shock wave would propagate along the cylinder, inducing $D+D$ fusion with an explosion of (in principle) unlimited power for a long enough tube. The essential step in the project was the calculation of the energy balance. The main source of energy loss was the bremsstrahlung — the production of γ-rays in electron-ion collisions, since γ-rays leave the system. In the course of propagation in the cylinder, γ-rays undergo Compton scattering. Because the bremsstrahlung spectrum is softer than the electron spectrum, the energy of γ's increases at Compton scattering. (Sometime this process is called inverse Compton scattering.) The calculation of this increase of energy loss due to Compton scattering and the calculation of the energy balance taking into account this effect was the main task of the Pomeranchuk group (in collaboration with the Zeldovich group). Much later, I read in Rhodes' book,[4] that at a conference at Los Alamos in April 1946, where Teller presented the results of his calculations of the Classical Super,

Bethe made a remark that the inverse Compton scattering of γ's (not taken into account by Teller) would result in a negative energy balance and that the hydrogen bomb of this type would not explode. In our calculations after hard work we came to the same conclusion — the energy balance was negative. Bethe's intuitive prediction was marvelous![a] Teller and his group came to the similar conclusions.

I met Hans for the first time in 1994, when he visited ITEP. He was invited into the ITEP Director's office, where there were also a few ITEP physicists present, including myself. According to standard procedure, ITEP Director Prof. I. V. Chuvilo told the guest about the main investigations proceeding in the Institute, and suggested the program of his visit: first to see this installation, then that and so on. But Bethe refused to follow this program and instead he said: "I would like to have a conversation with the ITEP theoreticians and first of all I would like to know what they are doing in QCD." So, we went to my office and for about two hours, and I explained to Hans our results in QCD: the calculations of baryon masses, based on the dominant role of spontaneous violation of chiral symmetry in a QCD vacuum, the calculations of baryon magnetic moments etc. Hans listened very attentively, posed questions and it was clear that he was keen to understand the subject more fully. Only once, after about one hour of conversation, he said: "I would like to rest for 10 minutes." I left the room, came back after 10 minutes and the conversation continued. (Hans was 87 at that time!)

The Necessity of Nuclear Power

Bethe's interest in nuclear power arose almost naturally; since the 1930's he had been one of the greatest authorities on nuclear physics. To illustrate this it may suffice to recall his reviews on theoretical nuclear dynamics,[7] his paper on diffraction scattering of nucleons on nuclei[8] and his lectures on nuclear theory.[9] In 1942–1946 Bethe was the Head of the Theoretical Division at Los Alamos. In 1956–1957 Bethe participated in experimental work, when the inelastic cross sections of fission spectrum neutrons were measured in various elements.[10] These data are important for construction of nuclear power reactors, especially of the fast neutrons breeder reactors.

Since the 1970's Bethe had been greatly concerned with global energy problems, especially with the balance of energy production and consumption

---

[a]This story is described in more details in Refs. 5, 6.

in the world in future. In two papers[11,12] published in 1976 and 1977 he discussed these problems. Bethe expected that the consumption of energy would steadily increase in the following decades, but the production of oil, the main source of energy, would drop around the year 2000 and the oil (and gas) prices would increase dramatically. Today we know that his prediction about rising oil and gas prices was correct; we also know, (and he did not discuss this in his early papers) that some developing countries like China, India and Brazil have joined the club of the main energy consumers. Bethe considered two ways of avoiding the future energy crises.

The first one was energy conservation — the improvement of efficiency with which the energy is utilized. He stressed the necessity of great efforts towards energy conservation: besides technical and industrial progress, public education and tax or other incentives were needed.[10] However, in this way only a temporary solution (for the next 10–20 years) of the problem might be achieved: in the long term, energy conservation alone would not help very much.

The second approach was the intensive use of new sources of energy. Among these the conversion of coal into synthetic gas or oil, nuclear fusion and solar power were considered. Bethe was sceptical about the extensive use of coal, especially in Europe, where coal resources are restricted and deep mining is needed. The use of nuclear fusion requires the solution of many hard engineering problems. So, Bethe expected[11] that, even in the best case scenario, fusion would contribute only a few per cent of the US's power supply by 2020. Solar power was considered in more details in[10] and the conclusion was: solar power is likely to remain extremely expensive in the US, Europe and Japan, but "it might be just the right thing for certain developing countries, which have large desert areas and rather modest total energy needs." Bethe concluded that all alternative sources of energy, including wind and biological sources, should be investigated and developed, but that these alone could not solve future energy problems.

Therefore, the final conclusion was that there was no alternative to nuclear power. "Nuclear power is a *necessity* not merely an option. A necessity if we want to make a smooth transition from our present oil- and gas-based economies to the post-oil world."[12]

Advantages and Problems of Nuclear Energy

Bethe emphasized that the main problem of a long-term future of nuclear energy based on breeder reactors was how to guarantee the non-proliferation of nuclear weapons. He shared the view[13] that this goal could not be achieved, if the plutonium or weapon-grade uranium producing facilities, like reprocessing or isotope separation plants, were widely distributed across the world. Bethe believed, that the widespread use of nuclear energy could only be reconciled with non-proliferation of nuclear weapons, if the fuel for nuclear reactors were not only uranium, but also thorium. He supported T. B. Taylor and H. A. Feiverson's proposals[13] for nuclear energy which advocated that nuclear power plants should be of two types. The first type are power plants with thermal reactors using ^{233}U, ^{238}U and thorium as a fuel. The moderator in such reactors should be heavy water. As the most promising heavy water reactors Bethe considered Canadian reactors CANDU.[b] In these one could expect a high conversion coefficient (the ratio of newly formed fission elements — ^{233}U and ^{239}Pu, ^{241}Pu to the burned ones) which was close to 1. The ratio of ^{233}U to ^{238}U would have to be chosen in such a way that after chemical separation of uranium it would not be suitable to manufacture nuclear weapons. Also, the concentration of plutonium should be much smaller than that of ^{233}U. The nuclear power plants of the second type are those with breeder reactors on ^{238}U and plutonium. The blankets of breeders would be made of thorium so as to produce ^{233}U for thermal reactors. The plutonium produced in thermal reactors would be fed back to breeder reactors. The number of fast breeders should be much smaller than the number of thermal reactors.

The problem of nonproliferation of nuclear weapons was weaker in the two-component scheme of nuclear energy plants, as proposed by Bethe, than in the case of energy production by fast breeder reactors. The number of breeder reactors in the two-component scheme is relatively small. They should be under international management and located in countries with great internal stability. The plants for reprocessing breeder blankets should be located in close proximity to them. Only the thermal reactors would be sold in international trade. Of course, as mentioned by Bethe, in order to prevent proliferation, some international agreement was necessary and, therefore, some restrictions to national sovereignty. But this was the price to be paid for world peace.

---

[b]In 1995 I participated at the E. Wigner Memorial Meeting organized by APS, where Bethe presented his recollections on Wigner. There was also a talk "Wigner as nuclear engineer" given by Alvin Weinberg. After this talk I gave a remark about the development of nuclear reactor theory in the USSR (stressing that it was independent from those of US) and about construction of heavy water reactors in USSR. After the meeting I had a conversation with Bethe — he did not know about heavy water reactors in the USSR. (The description of the development of heavy water reactors in USSR is given in Ref. 14.)

Bethe expected that nuclear energy formed in this way would cover all energy consumption of the world for thousands of years, and electric energy produced by nuclear power stations would be cheaper than that produced by using fossil fuel. (It must be mentioned that the last conclusion was based on the economic calculations performed before the Three Mile Island and Chernobyl accidents. After those, safety requirements for nuclear power stations were tightened, particularly in the US, where additional safety provisions resulted in a three-fold increase of costs. At the same time the price of fossil fuel also rose sharply, and a completely new economic analysis would be required now.) In carrying out the proposed nuclear energy program, the problems of chemical separation of the fuel containing thorium, uranium and plutonium as well as engineering problems have to be solved. One of the most difficult technical problems is the remote handling of fuel. The reason is that in the fuel, containing ^{233}U in the reactor core, due to $(n, 2n)$ reaction of fast neutrons, ^{232}U is formed and the products of a radiative chain starting from ^{232}U are highly radioactive. The flip side of this coin is that the high degree of radioactivity of the fuel provides a safeguard against its theft.

The other serious problems arising from widespread use of nuclear energy are the problem of waste disposal, and the release of radioactivity by nuclear power stations, especially in the case of serious accidents in nuclear reactors.

In his discussion of the problem of waste disposal Bethe referred to the Report to the American Physical Society by the Study Group on Nuclear Fuel Cycles and Waste Management (APS Study Group).[15] He cited the general conclusion of the Report: "Effective long-term isolation for spent fuel, high-level or transuranic wastes *can* be achieved by geological emplacement ... Many waste repository sites with satisfactory hydrogeology can be identified in continental US in a variety of geological formations. Bedded salt can be a satisfactory medium for a repository, but certain other rock types, notably granite and possibly shale, could offer even greater long-term advantages. Irrespective of the time scale, adopted for reprocessing, *two* geological demonstration facilities in *different* media should be completed." (emphasis in italics by Bethe). The wastes had to be solidified, then the solid wastes should be fused with borosilicate glass and put into steel cylinders. It is hard to imagine how the radioactive material could get out into the environment from the repository at a depth of 500 m below the ground after such treatment or how terrorists could penetrate there.

In some other countries (France, Russia) the decisions taken about waste disposal are close to the recommendations presented in the Report; in the US, however, the final decision about waste depository has still not been made. It must be mentioned that in recent years the proposal of transmutation by accelerators of radioactive transuranium elements into nonradioactive ones has been under discussion and corresponding experiments are under way. Bethe had a keen interest in this proposal and supported the construction of accelerators for this purpose.

In the discussion of the problem of radioactivity release by nuclear power plants Bethe remarked first that the threat posed by the routine operations of nuclear power plants through radioactive outflow was negligible in comparison with natural radioactivity, even for people permanently residing close to such a plant. The main danger arises in cases of serious accidents in reactors. The possibility and the consequences of such accidents were studied in the 1975 US Atomic Energy Commission Reactor Safety Study (the results were presented in the so-called Rasmussen report) and by the American Physical Society's Study Group on Light Water Reactor Safety. Later, in 1986, after the Chernobyl accident a special American panel was formed for its study. Hans Bethe participated in the APS Study Group and in the panel.

The Rasmussen report considered a nuclear reactor core meltdown accident followed by the release of radioactivity into the atmosphere, in the case when the containment failed to keep the radioactivity inside. The report estimated the probability of such an event as 1/200,000 per 1000 megawatts electrical reactor year. Of course, this estimate was purely theoretical, because no such accident has ever happend. The number of fatalities from radiation leak was estimated as 300. The APS study did not analyse the probability of such an accident, but by accounting the delayed cancers, came to much higher estimation of fatalities — about 10,000. Even if we take the APS estimation, remarks Bethe,[16] the number of fatalities from the accident — about 5 per year for the whole US — is not very high in comparison with other types of non-nuclear accidents. Naturally, Bethe shared the important conclusions of the APS Study Group[17]: (1) the necessity of improvement of containment design; (2) mitigation of accident consequences.

The main points of the report by the US panel that studied the Chernobyl accident were presented by Bethe.[18] Bethe stressed that "Chernobyl-type reactor design is seriously flawed." The reactor was inherently unstable as a physical system: the power of the reactor increased with increasing temperature or the content of steam in the cooling water. This was the main cause of this catastrophic accident. (I would like to mention, that this assessment completely concides with mine.[5,6]) The Chernobyl reactor did not have a containment. Therefore, the blow-up of the reactor immediately resulted in the crash of the roof of the reactor building and radioactivity was released into the atmosphere. Both these defects are absent in US light-water reactors. Moreover, the containments in US reactors are equipped with water sprays, which cool down the steam and radioactive products, condensing them within the containment. The most dangerous fission products — iodine and cesium (in the form of cesium iodide) are highly soluble in water, ejected by sprays. For these reasons, remarked Bethe, the amount of iodine ejected into the atmosphere during the Three Mile Island accident was relatively modest: only about one-millionth of the amount ejected at Chernobyl. No amounts of cesium or strontium were detected at the Three Mile Island accident. Bethe concluded that the likelihood of a seriously damaging accident was clearly much smaller for US light-water reactors than for the Chernobyl-type. But both accidents at Chernobyl and the less serious one at Three Mile Island indicated "that continued concern for safety is maintained over the full lifetime of US plants"[17] (I add: and over the whole world).

I am thankful to G. Brown for information about Bethe's activity in nuclear energy and to U. Meissner for his hospitality at Bonn University, where this paper was written.

This work was supported in part by US CRDF Cooperative Grant Program, Project RUP2-2621-MO-04, RFBR grant 03-02-16209 and the funds from EC to the project "Study of Strongly Interacting Matter" under contract 2004 No R113-CT-2004-506078.

References

1. H. A. Bethe, *Ann. d. Phys.* **3**, 133–208 (1929).
2. H. A. Bethe, *Ann. d. Phys.* **5**, 325–400 (1930).
3. H. A. Bethe and R. E. Peierls, *Proc. Roy. Soc. A* **148**, 146 (1935).
4. R. Rhodes, *Dark Sun*, Simon & Schuster, New York, 1995, p. 254.
5. B. L. Ioffe, *Osobo Secretnoe Zadanie, Novy Mir Magazine*, **5**, 144–155 (1999); **6**, 161–172 (1999) in Russian; English translation in: *Handbook of QCD*, ed. M. Shifman, v. 1, pp. 18–52, World Scientific, 2001.
6. B. L. Ioffe, *Without Retouching*, Phasis, 2004 (in Russian).
7. H. A. Bethe and R. F. Bacher, *Rev. Mod. Phys.* **8**, 193 (1936); H. A. Bethe, *Rev. Mod. Phys.* **9**, 71 (1937).
8. H. A. Bethe and G. Placzek, *Phys. Rev.* **57**, 1075A (1940).
9. H. A. Bethe, *Elementary Nuclear Theory*, John Wiley, New York, 1947; H. A. Bethe and P. Morrison, *Elementary Nuclear Theory*, 2nd ed., John Wiley, New York, 1956.

10. H. A. Bethe, J. R. Beyster and R. E. Carter, *J. Nuclear Energy* **3**, 207, 273 (1956); **4**, 3, 147 (1957).
11. H. A. Bethe, *Sci. Am.* **234**, 21 (1976).
12. H. A. Bethe, "The Debate on Nuclear Power," in *Nuclear Power and Its Fuel Cycle*, v. 7, pp. 3–17 (IAEA), Vienna, 1977.
13. H. A. Feiveson and T. B. Taylor, *Bull. Atomic Scientists*, December 1976, p. 14.
14. B. L. Ioffe and O. V. Shvedov, *Atomnaya Energia* **86**, 310 (1999).
15. Report to APS by the study group of nuclear fuel cycles and waste management, *Rev. Mod. Phys.* **50**, S1-185, January 1978, part II.
16. H. A. Bethe, *Bull. Atomic Scientists*, September 1975, p. 40.
17. F. von Hippel, *Bull. Atomic Scientists*, September 1975, p. 37.
18. H. A. Bethe, *Bull. Atomic Scientists*, December 1986, p. 45.

In Memoriam: Hans Bethe[*]

Richard L. Garwin and Frank von Hippel

Hans Bethe, who died on March 6 at the age of 98, was exemplary as a scientist; a citizen-advocate seeking to stem the arms race; and an individual of warmth, generosity, tenacity, and modest habits. Bethe made major contributions to several areas of physics during his academic career. He earned a Nobel Prize in 1967 for his research into how the sun generates its energy by converting hydrogen to helium using carbon as a nuclear catalyst. A few years later, he made central contributions to the secret US World War II nuclear-weapon development programs (the "Manhattan Project").

As did some other talented physicist-refugees from Europe, Bethe joined the Manhattan Project out of fear that Nazi Germany might be developing a nuclear bomb. After contributing to the development of radar at the wartime MIT Radiation Laboratory, Bethe moved to Los Alamos where he was chosen to lead the effort to assess mathematically the explosive properties of the evolving designs of the Hiroshima and Nagasaki bombs. Bethe was universally respected in the fractious physics community, and his calculations were definitive. For example, when Edward Teller raised the concern in 1942 that a nuclear explosion might ignite a fusion chain reaction in the atmosphere or ocean, Bethe showed clearly that this was impossible.[1]

At the end of World War II, Bethe returned to Cornell and became the inspiring figure at the core of a first-rate physics department. His own students included Richard Feynman and Freeman Dyson, both of whom made central contributions to the development of the quantum field theory of electromagnetism as well as becoming well known for their observations on technology

---

[*]Article originally published in *Arms Control Today*, Vol. 35, No. 3, April 2005, pp. 44–46. Reproduction rights granted by the Arms Control Association, Washington, D.C.

and life. At Cornell as at Los Alamos, Bethe showed the sweetness and generosity of character that was his personal hallmark. He always had time for younger physicists, and one always left a session with him encouraged.

After the war, with the veil of secrecy partially lifted, Bethe began to contribute in public commentary as well as in internal governmental councils to the national debate over nuclear weapons policy. In December 1945, he met in Washington with other former Manhattan Project scientists to found the Federation of American Scientists to urge international control of nuclear energy and "to disseminate those facts necessary for intelligent conclusions concerning the social implications of new knowledge in science."[2] Over the following decades, Bethe became the most respected of the scientists explaining the technical facts to the public, especially in a series of *Scientific American* articles written at crucial junctures. The issues with which he was most engaged were:

▶ The decision to develop the hydrogen bomb in 1950.
▶ The global campaign in the late 1950's and early 1960's to end nuclear testing.
▶ The effort over two decades to prevent a destabilizing offensive-defensive arms race.

The Hydrogen Bomb Decision

With some other scientific leaders of the World War II bomb project, including the members of the General Advisory Committee of the Atomic Energy Commission, Bethe advised as an insider against a crash program to develop a hydrogen bomb as a response to the first Soviet nuclear test in August 1949. President Harry Truman rejected this advice and announced his decision to go ahead on January 31, 1950. After the announcement, Bethe used the pages of *Scientific American* to explain to the American public his fundamental objection to such a weapon:

> I believe the most important question is the moral one: Can we, who have always insisted on morality and human decency between nations as well as inside our own country, introduce this weapon of total annihilation into the world? The usual argument, heard in the frantic week before the President's decision and frequently since, is that we are fighting against a country which denies all the human values we cherish, and that any weapon, however terrible, must be used to prevent that country and its creed from dominating

the world. It is argued that it would be better for us to lose our lives than our liberty; and with the view I personally agree. But I believe that this is not the choice facing us here; I believe that in a war fought with hydrogen bombs we would lose not only many lives but all our liberties and human values as well.[3]

Having lost the debate, Bethe nevertheless returned to Los Alamos as a consultant and, after failing to prove that it would be infeasible to build a thermonuclear bomb, contributed to the design effort. Indeed, following the proposal by Teller and Stanislaw Ulam in March 1951 that the X-ray radiation from a fission explosion could be used to implode and ignite fusion fuel, Bethe led the theoretical program to develop thermonuclear weapons, as he had done with the fission weapon. Bethe later explained that, "[i]f I didn't work on the bomb, somebody else would — and I had the thought that if I were around Los Alamos, I might still be a force for disarmament. It seemed quite logical. But sometimes I wish I were more consistent an idealist."[4]

Teller, by contrast, had been obsessed with the need to develop the H-bomb ever since Enrico Fermi suggested the possibility to him in 1941.[5] After his dream was finally realized, Teller was lionized by the right as "the father of the H-bomb" and became a leading proponent of the need to stay ahead of the Soviets in the arms race and for the deployment of ballistic missile defenses. Teller and Bethe, once close friends, were fated to be on opposite sides of arms control debates for the rest of their lives. Teller died in September 2003 at the age of 95.

The Test Ban

The international campaign for a nuclear test ban began in 1954 after a US hydrogen bomb test showered a Japanese fishing boat, *The Lucky Dragon*, with radioactive fallout that sickened the crew and ultimately killed one sailor. Bethe was able to galvanize White House consideration of the issue in 1957 as a member of President Dwight Eisenhower's new Science Advisory Committee. Eisenhower had created the panel after the Soviet Union shocked Washington by launching the first artificial earth satellite. Following Bethe's suggestion, an interagency panel was set up under his chairmanship to assess the verifiability of a test ban and whether a test ban would be to the benefit of the United States. The panel came up with the first design for a verification system, laying the basis for an international conference of experts in 1958 in which Bethe participated. The conference agreed on a seismic system consisting of 180 stations and concluded that underground tests with yields above five kilotons could be detected.[6]

The verification system was torpedoed, however, by an ingenious evasion scenario concocted by Teller and colleagues, who argued that the Soviet Union would be able to build and conceal underground spherical cavities 1000 feet in diameter to muffle underground nuclear explosions up to 300 kilotons in yield. On-site inspections therefore would be required to distinguish muffled underground explosions from small earthquakes. Ultimately, the United States insisted on the right to have more on-site inspections annually than the Soviet Union was willing to accept, and the result was that only a Partial Test Ban Treaty could be achieved in 1963, banning nuclear tests everywhere except underground.[7]

Missile Defense

As soon as Sputnik demonstrated the feasibility of intercontinental ballistic missiles, the US military began to propose systems to shoot them down as a complement to the extensive continental air defense that had already been deployed against Soviet bombers. The president's science advisers showed the inadequacies of one design after another until, in 1967, Richard Nixon announced that he would make a campaign issue out of the fact that the President Lyndon B. Johnson had not deployed a missile defense. Johnson responded by ordering the deployment of a "thin" missile defense called "Sentinel," nominally designed to defend against the missiles that China was then developing (but only tested 11 years later).

The Pentagon erred, however, when it decided to site nuclear-armed antimissile rockets in the suburbs of major US cities, including Chicago, Boston, and Seattle. Instead of accepting the protection gratefully, the suburbs were alarmed by the possibility that one of the nearby nuclear warheads might explode accidentally and rose up in an early manifestation of the ("not in my backyard") phenomenon. The ruckus created an audience in Congress for arguments against the defense system.

Fortunately, Bethe had already co-authored an article explaining how any country that could develop ICBMs could also neutralize the proposed system by, for example, adding decoys to their payloads that would overwhelm the defense.[8] This paper educated hundreds of physicists, who then used it to educate their communities and Congress on the issues. As a result, congressional support for the system steadily eroded until then-President Nixon was forced to negotiate the US-Soviet Anti-Ballistic Missile Treaty of 1972, which limited each country to two (later one) interceptor sites.[9] The United States shut down its permitted site in 1976 after only a few months of operation.

After President Ronald Reagan announced his Strategic Defense initiative in 1983, Bethe and collaborators again explained the many Soviet options to defeat the proposed system, which this time was to include a constellation of orbiting lasers.[10] Once again, their technical critique contributed significantly to the erosion of congressional support for what became derisively known as "Star Wars," and the development program lost most of its funding.

No New Nuclear Weapons

After the end of the Cold War, in 1995, at the age of 88 and on the occasion of the 50th anniversary of Hiroshima, Bethe decided that it was time to call on the world's weapons scientists to help end what he had helped begin. He used Los Alamos as a platform to address scientists there directly as well as scientists around the world through the press:

> Looking back at the half-century since [their creation], I feel the most intense relief that these weapons have not been used since World War II, mixed with the horror that tens of thousands of such weapons have been built since that time — one hundred times more than any of us at Los Alamos could ever have imagined.
>
> Today we are rightly in an era of disarmament and dismantlement of nuclear weapons... But in some countries nuclear development still continues. Whether and when the various Nations of the world can agree to stop this is uncertain. But individual scientists can still influence this process by withholding their skills.
>
> Accordingly, I call on all scientists in all countries to cease and desist from work creating, developing, improving and manufacturing further nuclear weapons — and, for that matter, other weapons of potential mass destruction such as chemical and biological weapons.[11]

References

1. R. Rhodes, *The Making of the Atomic Bomb*, Simon & Schuster, New York, 1987, p. 419. See H. A. Bethe, "Can Air or Water Be Exploded?," *Bull. Atomic Scientists*, March 15, 1946; Hans Bethe, "Ultimate Catastrophe?," *Bull. Atomic Scientists*, June 1976, p. 36.

2. A. K. Smith, *A Peril and a Hope: The Scientists' Movement in America, 1945–1947*, rev. ed., MIT Press, Cambridge, MA, 1971, p. 236.
3. H. A. Bethe, "The Hydrogen Bomb: II," *Sci. Am.*, April 1950.
4. S. S. Schweber, *In the Shadow of the Bomb: Oppenheimer, Bethe, and the Moral Responsibility of the Scientist*, Princeton University Press, Princeton, NJ, 2000, p. 166.
5. R. Rhodes, *The Making of the Atomic Bomb*, p. 374.
6. S. S. Schweber, *In the Shadow of the Bomb*, pp. 175–177.
7. For a fuller description of President John Kennedy's decision to seek only a Limited Test Ban Treaty, see Wade Boese and Daryl Kimball, "The Limited Test Ban Treaty Turns 40," *Arms Control Today*, October 2003, pp. 37–38.
8. R. Garwin and H. A. Bethe, "Anti-Ballistic-Missile Systems," *Sci. Am.*, March 1968.
9. J. Primack and F. von Hippel, *Advice and Dissent: Scientists in the Political Arena*, Basic Books, New York, 1974.
10. H. A. Bethe *et al.*, "Space-Based Ballistic-Missile Defense," *Sci. Am.*, October 1984.
11. S. S. Schweber, *In the Shadow of the Bomb*, p. 171.

Obituary: Hans A. Bethe*
Kurt Gottfried

Hans Bethe, who died on 6 March 2005, was the last of the brilliant young theorists who entered physics right after quantum mechanics was discovered. In 1926, at the age of 20, he joined Arnold Sommerfeld's seminar in Munich just as Erwin Schrödinger's papers began to appear. He quickly demonstrated exceptional power and ingenuity. By 1931 his publication list included three classics: the spectrum of an atom embedded in a lattice, one of the first applications of group theory to quantum mechanics; a complete solution of the 1D Heisenberg ferromagnet using the famous Bethe Ansatz; and the first detailed quantum theory of energy loss suffered by charged particles traversing matter.

Because his mother was Jewish at birth, Bethe was dismissed from his post at Tübingen when the Nazis came to power. After two highly productive years in England, he moved to Cornell University in 1935, where he was to remain for the rest of his life.

Bethe had an unequalled ability to synthesize and elucidate complex newly developed knowledge. This was first demonstrated in the 1933 *Handbuch der Physik* by a long article on solid-state physics, and another on one- and two-electron atoms; and a few years later in three issues of *Reviews of Modern Physics*, which became known as the Bethe Bible on nuclear physics.

This mastery of nuclear physics had two remarkable consequences. In 1938, Bethe discovered the carbon cycle, the intricate catalytic mechanism that turns hydrogen into helium in massive stars. In 1967, this work won the Nobel Prize for Physics — the first one to be awarded for a topic in astronomy.

---

*Reprinted with permission from *CERN Courier*, June 2005, p. 42. © 2005 CERN.

The second could not be more different: Bethe's leadership of the Theory Division at wartime Los Alamos. The phenomena relevant to nuclear explosions were so inaccessible to experiment that theory of all sorts was indispensable, and Bethe's intellectual powers and calm persona were needed to coordinate the stellar team that Robert Oppenheimer had assembled, consisting of people who had previously worked on whatever interested them, and usually alone.

After the war, Bethe worked intensely and simultaneously in two entirely different settings: at Cornell on pure academic physics, and as a senior advisor to — and critic of — the US government.

To an extent that was unique among the former leaders of the Manhattan Project, Bethe devoted great effort to what might appear to be contradictory ends: as a consultant to further weapons work and an opponent of such work, and as an advisor on US security policy and an opponent of central themes in this policy. This was because he held deep moral convictions and a strong pragmatic inclination.

From the start he was an outspoken advocate of arms control, and played a key part in establishing the atmospheric test ban. He publicly opposed developing the hydrogen bomb, but when it became known that such a device was possible he worked on it because he decided that the Soviets would soon have it. He worked on missile defence inside the government, concluded it would be both futile and counterproductive, and thereafter publicly opposed all attempts to deploy such systems. In a ceremony at Los Alamos on the 50th anniversary of Hiroshima, he called on scientists everywhere to desist from developing new nuclear weapons.

After the war, and thanks largely to Bethe, Cornell attracted some of the most talented physicists at Los Alamos — Richard Feynman and Robert Wilson, to name only the most famous. But Bethe always kept his own hands in front-line research until well over the age of 90. His first major post-war paper was his famous, rough-and-ready calculation of the Lamb shift, done on the train ride from the conference where Willis Lamb first announced that the Dirac equation did not account fully for the hydrogen spectrum. He then participated in virtuoso QED calculations with Feynman and their students.

During the 1950's and 1960's, he focused on nuclear matter, including the equation of state at high densities, which is important in astrophysics. After his retirement, he collaborated extensively with Gerald Brown at SUNY, Stony Brook, calling himself "Gerry's postdoc," and worked for nearly two decades on type II supernovae. After he became convinced that the solar-neutrino problem was not a fault of solar models, he wrote a landmark paper on the implications of neutrino oscillations, and important follow-on articles with John Bahcall.

Bethe was not only a truly outstanding scientist, but also a man of legendary candour and honesty. He was a teacher and mentor to generations of young physicists. It was instructive to see him handling reporters from the Cornell undergraduate newspaper as respectfully as the Washington press corps. And he had a great sense of humor. In 1931 he published a spoof of Arthur Eddington's claim that he had calculated the fine structure constant from first principles. Bethe and two other youngsters published a "calculation" of the absolute zero (in Centigrade units!) from the fine structure constant in *Naturwissenschaften*, and caused a scandal. In 1997, when World Scientific published a massive volume of selected papers, Bethe made sure this spoof was included.

When his death was announced on the front page of *The New York Times*, someone not at Cornell or in physics, but who knew him, wrote and asked, "Do they make them like that anymore?".

List of Publications of Hans A. Bethe

I. Books

1. *Elementary Nuclear Theory, A Short Course on Selected Topics*
   John Wiley, New York; Chapman and Hall, London, 1947

2. (with S.S. Schweber and F. de Hoffmann) *Mesons and Fields. Vol. I. Fields*
   Row, Peterson and Co., Evanston, IL, 1955

3. (with F. de Hoffmann) *Mesons and Fields. Vol. II. Mesons*
   Row, Peterson and Co., Evanston, IL, 1955

4. (with E.E. Salpeter) *Quantum Mechanics of One-and Two-Electron Atoms*
   Academic Press, New York, 1957

5. (with R.W. Jackiw) *Intermediate Quantum Mechanics*
   W.A. Benjamin, New York, 1964

6. *The Road from Los Alamos*
   Simon & Schuster, New York, 1991

7. *Selected Works of Hans A. Bethe*
   World Scientific, Singapore, 1997

8. (with G.E. Brown and C.-H. Lee) *Formation and Evolution of Black Holes in the Galaxy: Selected Papers with Commentary*
   World Scientific, Singapore, 2003

II. Papers

1. (with A. Bethe and Y. Terada) Versuche zur Theorie der Dialyse (Experiments Relating to the Theory of Dialysis)
   Zeitschrift f. Physik. Chemie, **112**, 250–69 (1924)

2. Theorie der Beugung von Elektronen an Kristallen (Theory of the Diffraction of Electrons by Crystals)
   Ann. Phys. **87**, 55–129 (1928)

3. Über den Durchgang von Kathodenstrahlen durch gitterförmige elektrische Felder (Passage of Cathode Rays through Electric Fields formed by Grids)
   Z. Phys. **54**, 703–710 (1929)

4. Vergleich der Elektronenverteilung im Heliumgrundzustand nach verschiedenen Methoden (Comparison of the Distribution of Electrons in the Helium Ground State as Calculated by Different Methods)
   Z. Phys. **55**, 431–436 (1929)

5. Termaufspaltung in Kristallen (Splitting of Terms in Crystals)
   translated by Consultants Bureau, New York
   Ann. Phys. **3**, 133–208 (1929)

6. Berechnung der Elektronenaffinität des Wasserstoffs (Calculation of Electronic Affinity of Hydrogen)
   Z. Phys. **57**, 815–821 (1929)

7. Über die nichtstationäre Behandlung des Photoeffekts (Non-Stationary Treatment of the Photoelectric Effect)
   Ann. Phys. **4**, 443–449 (1930)

8. Zur Theorie des Durchgangs schneller Korpuskularstrahlen durch Materie (Theory of Passage of Swift Corpuscular Rays through Matter)
   Ann. Phys. **5**, 325–400 (1930)

9. Zur Theorie des Zeemaneffektes an den Salzen der seltenen Erden (Theory of the Zeeman Effect in the Salts of Rare Earth)
   Z. Phys. **60**, 218–233 (1930)

10. (with G. Beck and W. Riezler) On the Quantum Theory of the Temperature of Absolute Zero
    Die Naturwissenschaften **19**, 39 (1931)

11. Change of Resistance in Magnetic Fields
    Nature **127**, 336–337 (1931)

12. Zur Theorie der Metalle. I. Eigenwerte und Eigenfunktionen der linearen Atomkette (Theory of Metals. Part I. Eigenvalues and Eigenfunctions of the Linear Atomic Chain)
    Z. Phys. **71**, 205–226 (1931)

13. Bremsformel für Elektronen relativistischer Geschwindigkeit (Scattering of Electrons of Relativistic Velocity)
    Z. Phys. **76**, 293–299 (1932)

14. (with E. Fermi) Über die Wechselwirkung von zwei Elektronen (Interaction of Two Electrons)
    Z. Phys. **77**, 296–306 (1932)

15. Quantenmechanik der Ein und Zwei-Elektronenprobleme (Quantum Mechanics of One- and Two-Electron Problems)
    Handbuch der Physik, 24 Part I, Springer, Berlin, 1933

16. (with A. Sommerfeld) Elektronentheorie der Metalle (Electron Theory of Metals)
    Handbuch der Physik, 24 Part II, Springer, Berlin, 1933, pp. 333–622

17. (with H. Fröhlich) Magnetische Wechselwirkung der Metallelektronen. Zur Kritik der Theorie der Supraleitung von Frenkel (Magnetic Interaction of Metallic Electrons. About Frenkel's Criticism of the Theory of Superconductivity)
    Z. Phys. **85**, 389–397 (1933)

18. Theorie des Ferromagnetismus (Theory of Ferromagnetism)
    in: P.J.W. Debye (ed.), *Magnetismus*, Hirzel, Leipzig, 1933, pp. 74–81

19. Zur Kritik der Theorie der Supraleitung von R. Schachenmeier (About R. Schachenmeier's Criticism of the Theory of Superconductivity)
    Z. Phys. **90**, 674–679 (1934)

20. (with R. Peierls) Photoelectric Disintegration of the Diplon
    Int. Conf. on Phys., London, 93–94 (1934)

21. (with W. Heitler) On the Stopping of Fast Particles and on the Creation of Positive Electrons
    Proc. R. Soc. A **146**, 83–112 (1934)

22. The Influence of Screening on the Creation and Stopping of Electrons
    Proc. Cambridge Philos. Soc. **30**, 524–539 (1934)

23. (with R.E. Peierls) The "Neutrino"
    Nature **133**, 532–533 & 689–690 (1934)

24. (with A.H. Compton) Composition of Cosmic Rays
    Nature **134**, 734–735 (1934)

25. Quantitative Berechnung der Eigenfunktion von Metallelektronen (Quantitative Calculation of the Eigenfunction of Electrons in Metals)
    Helv. Phys. Acta **7**, Suppl. II, 18–23 (1934)

26. Memorandum on Cosmic Rays
    Carnegie Institution of Washington Yearbook 34, 333–335 (1935)

27. Ionization Power of a Neutrino with Magnetic Moment
    Proc. Cambridge Philos. Soc. **31**, 108–115 (1935)

28. (with R. Peierls) Quantum Theory of the Diplon
    Proc. R. Soc. A **148**, 146–156 (1935)

29. (with R. Peierls) The Scattering of Neutrons by Protons
    Proc. R. Soc. A **149**, 176–183 (1935)

30. On the Annihilation Radiation of Positrons
    Proc. R. Soc. A **150**, 129–141 (1935)

31. Statistical Theory of Superlattices
    Proc. R. Soc. A **150**, 552–575 (1935)

32. Masses of Light Atoms from Transmutation Data
    Phys. Rev. **47**, 633–634 (1935)

33. Theory of Disintegration of Nuclei by Neutrons
    Phys. Rev. **47**, 747–759 (1935)

34. (with D.F. Weekes and M.S. Livingston) A Method for the Determination of the Selective Absorption Regions of Slow Neutrons
    Phys. Rev. **49**, 471–473 (1936)

35. An Attempt to Calculate the Number of Energy Levels of a Heavy Nucleus
    Phys. Rev. **50**, 332–341 (1936)

36. Nuclear Radius and Many-Body Problem
    Phys. Rev. **50**, 977–979 (1936)

37. (with R.F. Bacher) Nuclear Physics. Part A. Stationary States of Nuclei
    Rev. Mod. Phys. **8**, 82–229 (1936)

38. Nuclear Physics. Part B. Nuclear Dynamics, Theoretical
    Rev. Mod. Phys. **9**, 69–244 (1937)

39. (with M.S. Livingston) Nuclear Physics. Part C. Nuclear Dynamics, Experimental
    Rev. Mod. Phys. **9**, 245–390 (1937)

40. (with M.E. Rose) Nuclear Spins and Magnetic Moments in the Hartree Model
    Phys. Rev. **51**, 205–213 (1937)

41. (with J.G. Hoffman and M.S. Livingston) Some Direct Evidence on the Magnetic Moment of the Neutron
    Phys. Rev. **51**, 214–215 (1937)

42. (with M.E. Rose) Kinetic Energy of Nuclei in the Hartree Model
    Phys. Rev. **51**, 283–285 (1937)

43. (with G. Placzek) Resonance Effects in Nuclear Processes
    Phys. Rev. **51**, 450–484 (1937)

44. (with F.H. Spedding) The Absorption Spectrum of $Tm_2(SO_4)_3{:}8H_2O$
    Phys. Rev. **52**, 454–455 (1937)

45. (with M.E. Rose) The Maximum Energy Obtainable from the Cyclotron
    Phys. Rev. **52**, 1254–1255 (1937)

46. The Oppenheimer–Phillips Process
    Phys. Rev. **53**, 39–50 (1938)

47. The Binding Energy of the Deuteron
    Phys. Rev. **53**, 313–314 (1938)

48. Order and Disorder in Alloys
    J. Appl. Phys. **9**, 244–251 (1938)

49. (with M.E. Rose and L.P. Smith) The Multiple Scattering of Electrons
    Proc. Am. Philos. Soc. **78**, 373–383 (1938)

50. Magnetic Moment of Li^7 in the Alpha-Particle Model
    Phys. Rev. **53**, 842 (1938)

51. (with E.J. Konopinski) The Theory of Excitation Functions on the Basis of the Many-Body Model
    Phys. Rev. **54**, 130–138 (1938)

52. (with C.L. Critchfield) The Formation of Deuterons by Proton Combination
    Phys. Rev. **54**, 248–254 (1938)

53. Coulomb Energy of Light Nuclei
    Phys. Rev. **54**, 436–439 (1938)

54. (with C.L. Critchfield) On the Formation of Deuterons by Proton Combination
    Phys. Rev. **54**, 862 (1938)

55. A Method for Treating Large Perturbations
    Phys. Rev. **54**, 955–967 (1938)

56. Energy Production in Stars
    Phys. Rev. **55**, 103 (1939)

57. (with M.E. Rose) On the Absence of Polarization in Electron Scattering
    Phys. Rev. **55**, 277–289 (1939)

58. Energy Production in Stars
    Phys. Rev. **55**, 434–456 (1939)

59. The Meson Theory of Nuclear Forces
    Phys. Rev. **55**, 1261–1263 (1939)

60. (with J.G. Kirkwood) Critical Behavior of Solid Solutions in the Order–Disorder Transformation
   J. Chem. Phys. **7**, 578–582 (1939)

61. (with W.J. Henderson) Evidence for Incorrect Assignment of the Supposed Si^{27} Radioactivity of 6.7-Minute Half-Life
   Phys. Rev. **56**, 1060–1061 (1939)

62. (with R.E. Marshak) The Physics of Stellar Interiors and Stellar Evolution
   Rep. Prog. Phys. **6**, 1–15 (1939)

63. (with F. Hoyle and R. Peierls) Interpretation of Beta-Disintegration Data
   Nature **143**, 200–201 (1939)

64. (with R.E. Marshak) The Generalized Thomas–Fermi Method as Applied to Stars
   Astrophys. J. **91**, 329–343 (1940)

65. The Meson Theory of Nuclear Forces. I. General Theory
   Phys. Rev. **57**, 260–272 (1940)

66. The Meson Theory of Nuclear Forces. Part II. Theory of the Deuteron
   Phys. Rev. **57**, 390–413 (1940)

67. (with S.A. Korff and G. Placzek) On the Interpretation of Neutron Measurements in Cosmic Radiation
   Phys. Rev. **57**, 573–587 (1940)

68. (with M.G. Holloway) Cross Section of the Reaction $N^{15}(p,\alpha)C^{12}$
   Phys. Rev. **57**, 747 (1940)

69. Recent Evidence on the Nuclear Reactions in the Carbon Cycle
   Astrophys. J. **92**, 118–121 (1940)

70. (with L.W. Nordheim) On the Theory of Meson Decay
   Phys. Rev. **57**, 998–1006 (1940)

71. A Continuum Theory of the Compound Nucleus
   Phys. Rev. **57**, 1125–1144 (1940)

72. (with G. Blanch, A.N. Lowan and R.E. Marshak) The Internal Temperature-Density Distribution of the Sun
   Astrophys. J. **94**, 37–45 (1941)

73. (with E. Teller) Deviations from Thermal Equilibrium in Shock Waves
   (Distributed by Ballistic Research Laboratories, Aberdeen, Maryland; later reproduced by Engineering Research Institute, University of Michigan), 1941

74. Energy Production in Stars
   Am. Sci. **30 (10)**, 243–264 (1942)

75. The Theory of Shock Waves for an Arbitrary Equation of State
    (Published originally for the Office of Scientific Research and Development, Report No. 545 (1942); this is a retyping of the report in Poulter Research Laboratories Library, No. G-320)

76. Theory of Diffraction by Small Holes
    Phys. Rev. **66**, 163–182 (1944)

77. Can Air Or Water Be Exploded?
    Bull. Atomic Scientists **1 (7)** (1946)

78. (with J.R. Oppenheimer) Reaction of Radiation on Electron Scattering and Heitler's Theory of Radiation Damping
    Phys. Rev. **70**, 451–458 (1946)

79. Multiple Scattering and the Mass of the Mesons
    Phys. Rev. **70**, 821–831 (1946)

80. (with F. Seitz) How Close Is the Danger?
    in: D. Masters and K. Way (eds.), *One World or None*, Latimer House, London, 1946–1947, pp. 92–101

81. (with H.S. Sack) German Scientists in Army Employment: II—A Protest
    Bull. Atomic Scientists **3 (2)**, 65, 67 (1947)

82. (with H.H. Barschall) Energy Sensitivity of Fast Neutron Counters
    Rev. Sci. Instrum. **18**, 147–149 (1947)

83. (with F.C. von der Lage) A Method for Obtaining Electronic Eigenfunctions and Eigenvalues in Solids with an Application to Sodium
    Phys. Rev. **71**, 612–622 (1947)

84. The Electromagnetic Shift of Energy Levels
    Phys. Rev. **72**, 339–341 (1947)

85. (with R.E. Marshak) On the Two-Meson Hypothesis
    Phys. Rev. **72**, 506–509 (1947)

86. (with M. Camac) The Scattering of High Energy Neutrons by Protons
    Phys. Rev. **73**, 191–196 (1948)

87. (with R.A. Alpher and G. Gamow) The Origin of Chemical Elements
    Phys. Rev. **73**, 803–804 (L) (1948)

88. Bemerkungen über die Wasserstoff-Eigenfunktionen in der Diracschen Theorie
    (Remarks Concerning the Hydrogen Eigenfunctions in Dirac's Theory)
    Zeits. Naturforschung **3A**, 470–477 (1948)

89. (with E.D. Courant) Electrostatic Deflection of a Betatron or Synchrotron Beam
    Rev. Sci. Instrum. **19**, 632–637 (1948)

90. (with C. Longmire) On the Experimental Value of the Fine Structure Constant
    Phys. Rev. **75**, 306–307 (1949)

91. Theory of the Effective Range in Nuclear Scattering
    Phys. Rev. **76**, 38–50 (1949)

92. (with U. Fano and P.R. Karr) Penetration and Diffusion of Hard X-Rays through Thick Barriers. I. The Approach to Spectral Equilibrium
    Phys. Rev. **76**, 538–540 (1949)

93. The Hydrogen Bomb
    Bull. Atomic Scientists **6 (4)**, 99–104 and 125 (1950)

94. (with L. Szilard, F. Seity and H.S. Brown) The Facts about the Hydrogen Bomb
    Bull. Atomic Scientists **6 (4)**, 106–109 (1950)

95. (with L.M. Brown and J.R. Stehn) Numerical Value of the Lamb Shift
    Phys. Rev. **77**, 370–374 (1950)

96. (with C. Longmire) The Effective Range of Nuclear Forces. II. Photo-Disintegration of the Deuteron
    Phys. Rev. **77**, 647–654 (1950)

97. (with J.S. Levinger) Dipole Transitions in the Nuclear Photo-Effect
    Phys. Rev. **78**, 115–129 (1950)

98. The Range-Energy Relation for Slow Alpha-Particles and Protons in Air
    Rev. Mod. Phys. **22**, 213–219 (1950)

99. The Hydrogen Bomb. II
    Sci. Am. **184**, 18–23 (1950)

100. (with L.M. Brown and M.C. Walske) Stopping Power of K-Electrons
    Phys. Rev. **79**, 413 (1950)

101. (with L. Tonks and H. Hurwitz, Jr.) Neutron Penetration and Slowing Down at Intermediate Distances through Medium and Heavy Nuclei
    Phys. Rev. **80**, 11–19 (1950)

102. (with R.L. Gluckstern) Neutron–Deuteron Scattering at High Energy
    Phys. Rev. **81**, 761–781 (1951)

103. (with H. Hurwitz, Jr.) Neutron Capture Cross Sections and Level Density
    Phys. Rev. **81**, 898 (1951)

104. (with M.C. Walske) Asymptotic Formula for Stopping Power of K-Electrons
    Phys. Rev. **83**, 457–458 (1951)

105. (with R.R. Wilson) Meson Scattering
    Phys. Rev. **83**, 690–692 (1951)

106. (with D.B. Beard) Field Corrections to Neutron–Proton Scattering in a New Mixed Meson Theory
Phys. Rev. **83**, 1106–1114 (1951)

107. (with J. Heidmann) Z-Dependence of the Cross Section for Photocapture by Nuclei
Phys. Rev. **84**, 274–281 (1951)

108. (with E.E. Salpeter) A Relativistic Equation for Bound-State Problems
Phys. Rev. **84**, 1232–1242 (1951)

109. (with H.C. Urey, C.S. Smith, W.P. Murphy, et al.) Eminent Scientists Give Their Views on American Visa Policy
Bull. Atomic Scientists **8 (7)**, 217–220 (1952)

110. (with J.S. Levinger) Neutron Yield from the Nuclear Photoeffect
Phys. Rev. **85**, 577–581 (1952)

111. (with S.T. Butler) A Proposed Test of the Nuclear Shell Model
Phys. Rev. **85**, 1045–1046 (1952)

112. (with F. Rohrlich) Small Angle Scattering of Light by a Coulomb Field
Phys. Rev. **86**, 10–16 (1952)

113. (with N. Austern) Angular Distribution of π^+-Production in n–p Collisions
Phys. Rev. **86**, 121–122 (1952)

114. (with H. Davies) Integral Cross Section for Bremsstrahlung and Pair Production
Phys. Rev. **87**, 156–157 (1952)

115. (with L.C. Maximon) Differential Cross Section for Bremsstrahlung and Pair Production
Phys. Rev. **87**, 156 (1952)

116. High Energy Phenomena (Course of lectures given at Los Alamos in the spring and summer of 1952)
January 8, 1953

117. (with M. Baranger and R. Feynman) Relativistic Correction to the Lamb Shift
Phys. Rev. **92**, 482–501 (1953)

118. Molière's Theory of Multiple Scattering
Phys. Rev. **89**, 1256–1266 (1953)

119. The Sign of the Phase Shifts in Meson–Nucleon Scattering
Phys. Rev. **90**, 994–995 (1953)

120. (with L.C. Maximon and F. Low) Bremsstrahlung at High Energies
Phys. Rev. **91**, 417–418 (1953)

121. What Holds the Nucleus Together?
     Sci. Am. **189**, 58–63 (1953)

122. Negotiations and Atomic Bombs
     Bull. Atomic Scientists **10 (1)**, 9–10 (1954)

123. (with L.C. Maximon) Theory of Bremsstrahlung and Pair Production. I. Differential Cross Section
     Phys. Rev. **93**, 768–784 (1954)

124. (with H. Davies and L.C. Maximon) Theory of Bremsstrahlung and Pair Production. II. Integral Cross Section for Pair Production
     Phys. Rev. **93**, 788–795 (1954)

125. Mesons and Nuclear Forces
     Phys. Today **7 (2)**, 5–11 (1954)

126. (with G. Breit) Ingoing Waves in Final State of Scattering Problems
     Phys. Rev. **93**, 888–890 (1954)

127. (with F.J. Dyson, M. Ross, E.E. Salpeter, S.S. Schweber, M.K. Sundaresan and W.M. Visscher) Meson–Nucleon Scattering in the Tamm–Dancoff Approximation
     Phys. Rev. **95**, 1644–1658 (1954)

128. (with F. de Hoffmann) Meson–Proton Scattering Phase Shift Analysis
     Phys. Rev. **95**, 1100–1101 (1954)

129. (with F. de Hoffmann, N. Metropolis and E.F. Alei) Pion-Hydrogen Phase Shift Analysis between 120 and 217 MeV
     Phys. Rev. **95**, 1586–1605 (1954)

130. (with F. de Hoffmann) Meson–Proton Scattering Phase Shift Analysis
     Phys. Rev. **96**, 1714 (1954)

131. Physik der pi-Mesonen
     in: E. Brüche and H. Franke (eds.), *Physikertagung Wiesbaden. Hauptvorträge der Jahrestagung 1955*, Physikverlag, Mosbach/Baden, 1955

132. Memorial Symposium Held in Honor of Enrico Fermi at the Washington Meeting of the American Physical Society, 29 April 1955
     Rev. Mod. Phys. **27**, 249 (1955)

133. Nuclear Many-Body Problem
     Phys. Rev. **103**, 1353–1390 (1956)

134. (with R. Dalitz and M. Sundaresan) A Singular Integral Equation in the Theory of Meson–Nucleon Scattering
     Proc. Cambridge Philos. Soc. **52**, 251–272 (1956)

135. General Introduction (Lecture at Amsterdam Nuclear Reactions Conf., July 2–7, 1956)
Physica **22**, 941–951 (1956)

136. Introduction to the Brueckner Theory (Lecture at Amsterdam Nuclear Reactions Conf., July 2–7, 1956)
Physica **22**, 987–993 (1956)

137. Reactor Safety and Oscillator Tests
APDA Report 117 (October 15, 1956)

138. (with J. Hamilton) Anti-Proton Annihilation
Nuovo Cimento **4**, 1–22 (1956)

139. (with J.R. Beyster and R.E. Carter) Inelastic Cross-Sections for Fission-Spectrum Neutrons. Part I
J. Nuclear Energy **3**, 207–223 (1956)

140. (with J.R. Beyster and R.E. Carter) Inelastic Cross-Sections for Fission-Spectrum Neutrons. Part II
J. Nuclear Energy **3**, 273–300 (1956)

141. (with J.R. Beyster and R.E. Carter) Inelastic Cross-Sections for Fission-Spectrum Neutrons. Part III
J. Nuclear Energy **4**, 3–25 (1957)

142. (with J.R. Beyster and R.E. Carter) Inelastic Cross-Sections for Fission-Spectrum Neutrons. Part IV
J. Nuclear Energy **4**, 147–163 (1957)

143. (with J. Goldstone) Effect of a Repulsive Core in the Theory of Complex Nuclei
Proc. R. Soc. A **238**, 551–567 (1957)

144. On the Doppler Effect in Fast Reactors
APDA Report **119**, 1957

145. (with E.E. Salpeter) Quantum Mechanics of One-and Two-Electron Systems
Handbuch der Physik **35**, 88–436 (1957)

146. (with B. Kivel and H. Mayer) Radiation from Hot Air. Part I. Theory of Nitric Oxide Absorption
Ann. Phys. (N.Y.) **2**, 57–80 (1957)

147. Scattering and Polarization of Protons by Nuclei
Ann. Phys. (N.Y.) **3**, 190–240 (1958)

148. Brighter Than a Thousand Suns
Bull. Atomic Scientists **14 (10)**, 426–428 (1958)

149. (with M.D. Adams) A Theory for the Ablation of the Glassy Materials
     J. Aero/Space Sci. **26**, 321–328 (1959)

150. (with K.S. Suh) Recoil Momentum Distribution in Electron Pair Production
     Phys. Rev. **115**, 672–677 (1959)

151. Fundamental Particles
     J. Franklin Inst. **268**, 501–507 (1959)

152. The Case for Ending Nuclear Tests
     Atlantic Monthly (August 1960), pp. 43–51

153. Soviet Test Plan Examined
     New York Times (March 29, 1960), 36

154. (with P. Carruthers) Role of the pi–pi Interaction in High Energy pi–Nucleon Interactions
     Phys. Rev. Lett. **4**, 536–539 (1960)

155. (with C.R. Schumacher) Usefulness of Polarized Targets and the Polarization Transfer Tensor in Reconstruction of the Nucleon–Nucleon Scattering Matrix
     Phys. Rev. **121**, 1534–1541 (1961)

156. Thinks Soviets Want Tests
     New York Times (July 2, 1961), E6

157. Interview with H.A. Bethe, conducted by Donald McDonald
     Bull. Atomic Scientists **18 (4)**, 25–27 (1962)

158. Disarmament and Strategy
     Bull. Atomic Scientists **18 (7)**, 14–22 (1962)

159. (with M. Leon) Negative Meson Absorption in Liquid Hydrogen
     Phys. Rev. **127**, 636–647 (1962)

160. (with T. Kinoshita) Behavior of Regge Poles in a Potential at Large Energy
     Phys. Rev. **128**, 1418–1424 (1962)

161. (with B.H. Brandow and A.G. Petschek) Reference Spectrum Method for Nuclear Matter
     Phys. Rev. **129**, 225–264 (1963)

162. (with A.L. Read and J. Orear) Exact Form for Scattering Amplitude in Regge Pole Theory
     Nuovo Cimento **29**, 1051–1058 (1963)

163. Derivation of the Brueckner Theory
     in: J.K. Percus (ed.), *The Many-Body Problem*, Interscience, New York, 1963, p. 31

164. Theory of the Fireball
    Los Alamos Report LA-3064, 1964

165. (with V.F. Weisskopf) Fermi Award Defended
    New York Times (March 23, 1964), 28

166. Note on High-Energy Proton–Proton Scattering
    Nuovo Cimento **33**, 1167–1172 (1964)

167. Disarmament and Strategic Stability
    Fourth Annual Dewey F. Fagerburg Memorial Lecture, Univ. of Michigan, 1964

168. Three-Body Correlations in Nuclear Matter
    Phys. Rev. **138**, B804–B822 (1965)

169. The Fireball in Air
    J. Quant. Spectrosc. Radiat. Transfer **5**, 9–12 (1965)

170. (with G.S. Janes, R.H. Levy and B.T. Feld) New Type of Accelerator for Heavy Ions
    Phys. Rev. **145**, 925–952 (1966)

171. Energy Production in the Sun
    in: T. Page and L.W. Page (eds.), *The Origin of the Solar System; Genesis of the Sun and Planets, and Life on Other Worlds*, Macmillan, New York, 1966, pp. 30–39

172. Shadow Scattering by Atoms
    in: A. de-Shalit, H. Feshbach, and L. Van Hove (eds.), Preludes in Theoretical Physics (in honor of V.F. Weisskopf), North-Holland, Amsterdam, 1966, pp. 240–249

173. J. Robert Oppenheimer, 1904–1967
    Bull. Atomic Scientists **23 (8)**, 3–6 (1967)

174. Oppenheimer: "Where He Was There Was Always Life and Excitement"
    Science **155**, 1080–1084 (1967)

175. (with P.J. Siemens) Shape of Heavy Nuclei
    Phys. Rev. Lett. **18**, 704–706 (1967)

176. Three-Body Correlations in Nuclear Matter. II
    Phys. Rev. **158**, 941–947 (1967)

177. (with R. Rajaraman) The Three-Body Problem in Nuclear Matter
    Rev. Mod. Phys. **39**, 745–770 (1967)

178. Theory of Nuclear Matter
    in: *Proc. Int. Conf. on Nuclear Structure*, Tokyo, Japan, September 1967, pp. 56–62

179. Energy Production in Stars (Nobel Lecture, December 11, 1967, Stockholm, Sweden)
    in: *Les Prix Nobel*, 1967

180. (with R.L. Garwin) Anti-Ballistic-Missile Systems
    Sci. Am. **218**, 21–31 (1968)

181. Thomas–Fermi Theory of Nuclei
    Phys. Rev. **167**, 879–907 (1968)

182. (with L.R.B. Elton) Charge Distribution in Nuclei
    Phys. Rev. Lett. **20**, 745–747 (1968)

183. (with J. Nemeth) A Simple Thomas–Fermi Calculation for Semi-infinite Nuclei
    Nucl. Phys. **A116**, 241–255 (1968)

184. (with J.W. Negele) Semi-analytical Theory of Muonic X-ray Levels
    Nucl. Phys. **A117**, 575–585 (1968)

185. J. Robert Oppenheimer
    Biographical Memoirs of Fellows of the Royal Society **14**, 391–416 (1968)

186. (with P.J. Siemens) Neutron Radius in Pb from the Analog State
    Phys. Lett. **27B**, 549–551 (1968)

187. The ABM, China and the Arms Race
    Bull. Atomic Scientists **25 (5)**, 41–44 (1969)

188. Washington Scene: Missiles and Antimissiles. Hard Point vs. City Defense
    Bull. Atomic Scientists **25 (6)**, 25–26 (1969)

189. M.I.T. Science Forum
    New York Times (February 15, 1969), 28

190. (with P.J. Siemens) Capture of K^- Mesons in the Nuclear Periphery
    Nucl. Phys. B **21**, 589–605 (1970)

191. (with G. Börner and K. Sato) Nuclei in Neutron Matter
    Astron. Astrophys. **7**, 279–288 (1970)

192. Introduction
    in: S. Fernbach and A. Taub, *Computers and Their Role*, Gordon and Breach, London, 1970, pp. 1–9

193. Disarmament Problems. Part 4 of "The Military Atom"
    Bull. Atomic Scientists **26 (6)**, 99–102 (1970)

194. Theory of Nuclear Matter
    Annu. Rev. Nucl. Sci. **21**, 93–244 (1971)

195. (with G. Baym and C.J. Pethick) Neutron Star Matter
    Nucl. Phys. A **175**, 225 (1971)

196. (with A. Molinari) Nonstatic Second Order Contribution to the Elastic Electron–Nucleus Scattering
    Ann. Phys. (N.Y.) **63**, 393–431 (1971)

197. Yalta: Lack of Communication on Bomb
    New York Times (February 28, 1971), E12

198. (with J.L. Fowler and P.J. Siemens) Atoms within Toroidal Nuclei
    Bull. Am. Phys. Soc. **16**, 1151 (1971)

199. Strutinsky's Energy Theorem
    in: H.P. Dürr (ed.), *Quanten und Felder*, Vieweg, Braunschweig, 1971, p. 237

200. (with M.B. Johnson and A. Molinari) Effective Two-Body Interaction for Equivalent Particles
    in: F. Calogero and C. Ciofi degli Atti (eds.), *The Nuclear Many-Body Problem*, Rome, 1972, p. 828

201. Note on Inverse Bremsstrahlung in a Strong Electromagnetic Field
    Los Alamos Report LA-5031, 1972

202. A Voice of Conscience is Stilled: IV Messages
    Bull. Atomic Scientists **29 (6)**, 8 (1973)

203. (with R.E. Rand, M.-R. Yearian and C.D. Buchanan) Comments on the Present Status of Elastic and Inelastic Magnetic Electron–Deuteron Scattering
    Phys. Rev. D **8**, 3229–3232 (1973)

204. Qualitative Theory of Pion Scattering by Nuclei
    Phys. Rev. Lett. **30**, 105–108 (1973)

205. (with V.R. Pandharipande) Variational Method for Dense Systems
    Phys. Rev. C **7**, 1312–1328 (1973)

206. Equation of State at Densities Greater than Nuclear Density
    in: C.J. Hansen (ed.), *Physics of Dense Matter*, Proceedings of IAU Symposium, 21–25 August 1972, Reidel, Dordrecht, Boston, 1974, pp. 27–46

207. (with G.C. Vlases) Diffusion in a Laser Heated Plasma Column
    Bull. Am. Phys. Soc. **19**, 932 (1974)

208. (with M.B. Johnson) Dense Baryon Matter Calculations with Realistic Potentials
    Nucl. Phys. A **230**, 1–58 (1974)

209. (with R.C. Malone and M.B. Johnson) Neutron Star Models with Realistic High-Density Equations of State
    Astrophys. J. **199**, 741–748 (1975)

210. Fuel for LWR, Breeders and Near-Breeders
    Ann. Nucl. Energy **2**, 763–766 (1975)

211. (with A. Moliari, M.B. Johnson and W.M. Albevier) Effective Two-Body Interaction in Simple Nuclear Spectra
    Nucl. Phys. A **239**, 45–73 (1975)

212. (with M.I. Sobel, P.J. Siemens and J.P. Bondorf) Shock Waves in Colliding Nuclei
    Nucl. Phys. **A251**, 502–529 (1975)

213. 32 Scientists Speak Out: "No Alternative to Nuclear Power"
    Bull. Atomic Scientists **31 (3)**, 4–5 (1975)

214. Nuclear Reactor Safety: "No Fundamental Change in the Situation"
    Bull. Atomic Scientists **31 (7)**, 40–41 (1975)

215. Report of Steering Review Committee of the American Physical Society's Study Group on Light Water Reactor Safety
    Bull. Atomic Scientists **(9)**, 35–37 (1975)

216. (with G.C. Vlases) Transverse Diffusion of Particles and Energy in a Laser Heated Plasma Column
    Phys. Fluids **18**, 982–990 (1975)

217. AEI Energy Policy
    Science **190**, 1152 (1975)

218. The Necessity of Fission Power
    Sci. Am. **234**, 21–31 (1976)

219. The Energy of the Stars
    Technology Review **78 (7)**, 58–63 (1976)

220. Ultimate Catastrophe?
    Bull. Atomic Scientists **32 (6)**, 36–37 (1976)

221. Testimony on the California Nuclear Initiative
    Energy **1**, 223–228 (1976)

222. The Case for Nuclear Power
    Skeptic 14 (July–August 1976), 22

223. Necessity of Fission Power
    Usp. Fiz. Nauk **120**, 455–475 (1976)

224. Influence of Gamow on Early Astrophysics and on Early Accelerators in Nuclear Physics, in *George Gamow Symposium*, eds. Eamon Harper, W.C. Parke, and David Anderson (ASP Conference Series, Vol. 129, 1977), p. 44ff

225. Energy Strategy
    Foreign Affairs **55**, 636–637 (1977)

226. (with J.H. Jacob) Diffusion of Fast Electrons in the Presence of an Electric Field
    Phys. Rev. A **16**, 1952–1963 (1977)

227. (with K. Adler, S.H. Bauer, et al.) Imprisoned Argentine Scientist
    Science **197**, 938 (1977)

228. Six Views on Atomic Energy: The Need for Nuclear Power
    Bull. Atomic Scientists **33 (3)**, 59–63 (1977)

229. Nuclear Power: A Safe Bet
    Bull. Atomic Scientists **33 (7)**, 55 (1977)

230. The Debate on Nuclear Power
    in: *Nuclear Power and Its Fuel Cycle 7*, International Atomic Energy Agency, Vienna, Austria, 1977, pp. 3–17

231. The Fusion Hybrid
    Nucl. News **21**, 41 (1978)

232. (with M.B. Johnson) Correlations and Self-Consistency in Pion Scattering (I)
    Nucl. Phys. **A305**, 418–460 (1978)

233. Interview with Hans A. Bethe conducted by Charles D. Pevsner
    Cornell Engineer **44 (11)**, 14–20 (1978)

234. The Fusion Hybrid
    Phys. Today **32 (5)**, 44–51 (1979)

235. (with G.E. Brown, J. Applegate and J.M. Lattimer) Equation of State in the Gravitational Collapse of Stars
    Nucl. Phys. A **324**, 487–533 (1979)

236. The Happy Thirties
    in: R.H. Stuewer (ed.), *Nuclear Physics in Retrospect — Proceedings of a Symposium on the 1930s*, May 1977, University of Minnesota Press, Minneapolis, 1979, pp. 9–31

237. Freeman Dyson's Views of His Times. Disturbing the Universe
    Phys. Today **32 (12)**, 51–52 (1979)

238. (with J.H. Applegate and G.E. Brown) Neutrino Emission from a Supernova Shock
    Astrophys. J. **241**, 343–354 (1980)

239. The Case for Coal and Nuclear Energy
    The Center Magazine **13** (May/June 1980), 14–22

240. The Lives of the Stars
    The Sciences **20 (8)**, 6–9 (1980)

241. Recollections of Solid-State Theory, 1926–1933
    Proc. Roy. Soc. A **371** 49–51 (1980)

242. (with R. Wilson and W. Hull) Scientists for SENSE
    Bull. Atomic Scientists **36 (7)**, 64 (1980)

243. Fulfilling Energy Strategy
    New York Times (July 1, 1980), A19

244. The Fusion Hybrid Reactor
    Sandia National Laboratories Report SAND 81-1265, 1981

245. Supernova Collapse and Explosion
    Bull. Atomic Scientists **26 (1)**, 17 (1981)

246. Meaningless Superiority
    Bull. Atomic Scientists **37 (8)**, 1, 4 (1981)

247. (with G.E. Brown and G. Baym) Supernova Theory
    Nucl. Phys. A **375**, 481–532 (1982)

248. Supernova Theory
    in: C.A. Barnes, D.D. Clayton and D.N. Schramm (eds.), *Essays in Nuclear Astrophysics*, Cambridge University Press, New York, 1982, pp. 439–466

249. (with K. Gottfried) The Five-Year Plan — A Loser Both Ways
    New York Times (June 10, 1982), A31

250. Comments on The History of the H-Bomb
    Los Alamos Science **3 (3)**, 42–53 (1982)

251. Supernova Shocks and Neutrino Diffusion
    in: M.J. Rees and R.J. Stoneham (eds.), *Supernovae: A Survey of Current Research*, D. Reidel, Amsterdam, 1982, pp. 35–52

252. The Arms Race
    Colloquium at Sandia National Laboratories, Albuquerque, NM, July 28, 1982

253. (with F.A. Long) The Value of a Freeze
    New York Times (September 22, 1982), A27

254. Energy Independence Is Possible
    E.L. Miller Lecture on Public Policy, Texas A&M University, October 8, 1982

255. (with E. Teller) The Nuclear Freeze: Will It Really Make Us Safer?
    Los Angeles Times (October 17, 1982)

256. (with A. Yahil and G.E. Brown) On the Neutrino Luminosity from a Type II Supernova
 Astrophys. J. **262**, L7–L10 (1982)

257. (with R.E. Bacher et al.) USC Recommendations
 Bull. Atomic Scientists **38 (6)**, 5 (1982)

258. The Inferiority Complex and American Nuclear Policy
 New York Review of Books **29 (10)**, 3 (1982)

259. Uranium Enrichment Technology
 Science **217**, 398 (1982)

260. (with K. Gottfried) Assessing Reagan's Doomsday Scenario
 The New York Times (April 11, 1982), E17

261. Hydrogen Bomb History
 Science **218**, 1270 (1982)

262. (with E.E. Salpeter) The 1983 Nobel Prize in Physics — Chandrasekhar and Fowler
 Science **222**, 881–883 (1983)

263. (with F.A. Long) The Freeze Referendum: What Next?
 Bull. Atomic Scientists **39 (2)**, 2–3 (1983)

264. (with K. Gottfried) Elusive Security: Do We Need More Nukes or Fewer?
 The Washington Post (February 6, 1983)

265. We Are Not Inferior to the Soviets
 in: K.W. Ford (ed.), Nuclear Weapons and Nuclear War, American Association of Physics Teachers, SUNY, Stony Brook, NY, 1983, pp. 9–18

266. (with G.E. Brown, J. Cooperstein and J.R. Wilson) A Simplified Equation of State Near Nuclear Density
 Nucl. Phys. A **403**, 625–648 (1983)

267. Nuclear Arms Education — The Urgent Need
 Phys. Today **36 (3)**, 120 (1983)

268. (with J. Cooperstein and G.E. Brown) Shock Propagation in Supernovae: Concept of Net Ram Pressure
 Nucl. Phys. A **429**, 527–555 (1984)

269. The Case against Laser ABM Development — Technology and Policy Views of Hans Bethe
 Laser Focus/Electro-Optics Magazine **20** (March 1984)

270. (with E. Teller) Face-Off on Nuclear Defense (Excerpts from a debate over Star Wars weaponry in November 1983 at the Kennedy School of Government at Harvard)
Technology Review (M.T.) **87 (3)**, 84 (1984)

271. (with R.L. Garwin, K. Gottfried and H.W. Kendall) Space-based Ballistic-Missile Defense
Sci. Am. **251**, 39–49 (1984)

272. (with D.M. Kerr and R.A. Jeffries) Obituary: Herman W. Hoerlin
Phys. Today **37 (12)**, 82–83 (1984)

273. (with K. Gottfried and H.W. Kendall) Reagan's Star Wars
New York Review of Books **31 (7)**, 47–52 (1984)

274. (R.L. Garwin, K. Gottfried, H.W. Kendall, C. Sagan and V.F. Weisskopf) Star Wars and the Scientists
Commentary **79 (3)**, 6–11 (1985)

275. (with G.E. Brown) How a Supernova Explodes
Sci. Am. **252**, 60–68 (1985)

276. (with J. Boutwell and R.L. Garwin) BMD Technologies and Concepts in the 1980s
Daedalus **114 (2)**, 53–71 (1985)

277. (with R.L. Garwin) Weapons in Space — Appendix A: New BMD Technologies
Daedalus **114 (3)**, 331–368 (1985)

278. (with R.S. McNamara) Reducing the Risk of Nuclear War
Atlantic Monthly **256 (1)**, 43–51 (1985)

279. (with N.E. Bradburry, R.L. Garwin, et al.) Tests Not Necessary
Bull. Atomic Scientists **41 (7)**, 11 (1985)

280. The Technological Imperative
Bull. Atomic Scientists **41 (7)**, 34–36 (1985)

281. (with J.R. Wilson) Revival of a Stalled Supernova Shock by Neutrino Heating
Astrophys. J. **295**, 14–23 (1985)

282. (with R.L. Garwin, K. Gottfried, et al.) Why Star Wars is Dangerous and Won't Work
New York Review of Books **32 (2)**, 26 (1985)

283. (with G. Yonas) Can Star Wars Make Us Safe?
Sci. Dig. **93 (9)**, 30 (1985)

284. Supernova Theory
in: A. Molinari and R. Ricci (eds.), *From Nuclei to Stars*, Soc. Italiana di Fisica, Bologna, 1985, pp. 181–209

285. Possible Explanation of the Solar-Neutrino Puzzle
Phys. Rev. Lett. **56**, 1305–1308 (1986)

286. (with J. Bardeen) Back to Science Advisers
New York Times (May 17, 1986), 27

287. (with H.J. Juretschke, A.F. Moodie and H.K. Wagenfeld)
Obituary: Paul P. Ewald
Phys. Today **39 (5)**, 101–104 (1986)

288. Supernova Theory
in: S.L. Shapiro and S.A. Teukolsky (eds.), *Highlights of Modern Astrophysics: Concepts and Controversies*, John Wiley, New York, 1986, pp. 45–83

289. U.S. Panel Assesses Chernobyl
Bull. Atomic Scientists **42 (10)**, 45–46 (1986)

290. (with W.T. Golden, R.F. Bacher, A.J. Goodpaster, E.R. Piore, I.I. Rabi and J.R. Kilian) Presidents Science Advisory Committee Revisited
Science, Technology and Human Values **11**, 5–19 (1986)

291. (with F. Seitz) A Chernobyl-Type Accident Can't Happen Here
New York Times (May 17, 1986), 27

292. (with G.E. Brown and J. Cooperstein) Convection in Supernova Theory
Astrophys. J. **322**, 201–205 (1987)

293. The Theory of Supernovae
in: T.T.S. Kuo and J. Speth (eds.), *Windsurfing the Fermi Sea*, Elsevier Science Publishers, Amsterdam, 1987, Vol. I, pp. 3–12

294. (with E. Baron, G.E. Brown, J. Cooperstein and S. Kahana) Type-II Supernovae from Prompt Explosions
Phys. Rev. Lett. **59**, 736–739 (1987)

295. Supernova Theory
in: N. Metropolis, D.M. Kerr, and G.-C. Rota (eds.), *New Directions in Physics*, Academic Press, Boston, 1987, pp. 235–256

296. (with G.E. Brown and J. Cooperstein) Stars of Strange Matter
Nucl. Phys. A **462 (4)**, 791–802 (1987)

297. Obituary: Richard Phillips Feynman (1918–1988)
Nature **332**, 588 (1988)

298. Nuclear Physics Needed for the Theory of Supernovae
Annu. Rev. Nucl. Part. Sci. **38**, 1–28 (1988)

299. (with G. Hildebrandt) Paul Peter Ewald
Biographical Memoirs of Fellows of the Royal Society **34**, 133–176 (1988)

300. (with R.L. Garwin, K. Gottfried, et al.) The Case of Mordechai Vanunu
    New York Review of Books **33 (10)**, 56 (1988)

301. Observations on the Development of the H-Bomb, published as Appendix II in the 1989 version of York, Herbert F., *The Advisors: Oppenheimer, Teller, and the Superbomb, with a Historical Essay by Hans A. Bethe*, Stanford Univ. Press, Stanford, CA, 1976, 1989

302. Chop Down the Nuclear Arsenals
    Bull. Atomic Scientists **45 (2)**, 11–15 (1989)

303. Better Missiles
    Bull. Atomic Scientists **45 (7)**, 44 (1989)

304. (with P.A.M. Dirac, W. Heisenberg, et al.) From a Life of Physics
    World Scientific, Singapore, 1989

305. Solar-Neutrino Experiments
    Phys. Rev. Lett. **63**, 837–839 (1989)

306. The Privilege of Being a Physicist: Victor F. Weisskopf
    Am. J. Phys. **57**, 957 (1989)

307. (with P. Pizzochero) Mass-Energy Relation for SN 1987A from Observations
    Astrophys. J. **350**, L33–L35 (1990)

308. Supernovae
    Phys. Today **43 (9)**, 24–27 (1990)

309. Nuclear History: Sakharov's H-Bomb
    Bull. Atomic Scientists **46 (8)**, 8–9 (1990)

310. The Genocidal Mentality
    Issues in Science and Technology **7 (1)**, 97–98 (1990)

311. (with J.N. Bahcall) A Solution of the Solar-Neutrino Problem
    Phys. Rev. Lett. **65**, 2233–2235 (1990)

312. How Supernova Shock Revival Was Revealed
    Phys. Today **43**, 43 (1990)

313. Supernova Mechanisms
    Rev. Mod. Phys. **62**, 801–866 (1990)

314. The Need for Nuclear Power
    California Institute of Technology Engineering & Science **54**, 28–35 (1991)

315. (with G.A. Gurzadyan) On the Possibility of Nuclear Processes in Atmospheres of Flare Stars
    Astrophysics and Space Science **183 (1)**, 1–10 (1991)

316. (with K. Gottfried and R.S. McNamara) The Nuclear Threat: A Proposal
     The New York Review of Books **38 (12)**, 48–50 (1991)

317. Chernobyl: It Can't Happen Here
     New York Times (May 2, 1991), A25

318. The Truth about Chernobyl
     New York Times (May 5, 1991), BR34

319. Binding Energies and Supernova Evolution — A Reply
     Phys. Today **44**, 15 (1991)

320. (with G.E. Brown and P.M. Pizzochero) The Hadron to Quark/Gluon Transition
     Phys. Lett. B **263**, 337–341 (1991)

321. (with J.N. Bahcall) Solar Neutrinos and the Mikheyev–Smirnov–Wolfenstein Theory
     Phys. Rev. D **44**, 2962–2969 (1991)

322. (with L. Rosen, N. Metropolis and G. Cowan) Labs Leap to Own Defence
     Bull. Atomic Scientists **48 (9)**, 45–46 (1992)

323. Solar Neutrinos
     Int. J. Mod. Phys. D **1**, 1–12 (1992)

324. (with J. Bjorksten and S.A. Korff) From Fireworks to Cosmic Rays, 1906 to 1940
     in: R.B. Mendell and A.I. Mincer (eds.), *Frontiers in Cosmic Physics, Annals of the New York Academy of Sciences* **655** (1992), p. 15ff

325. Supernova 1987A: An Empirical and Analytic Approach
     Astrophys. J. **412**, 192–202 (1993)

326. New Lives for Nuclear Labs
     New York Times (December 6, 1993), 17

327. Bethe on the German Bomb Project
     Bull. Atomic Scientists **49 (1)**, 53–54 (1993)

328. What Is the Future of Los Alamos?
     Bull. Atomic Scientists (1993)

329. Preservation of the Supernova Shock
     Astrophys. J. **419**, 197–199 (1993)

330. The Farm Hall Transcripts
     Am. J. Phys. (1993)

331. Life and Physics with Viki
"Achievements in Physics: The International Community of Physicists," honoring the lifetime achievements of V.F. Weisskopf, September 1993, by the American Institute of Physics

332. Genius in the Shadows
Reviewed by H.A. Bethe
Phys. Today (1993)

333. (with J.N. Bahcall) Do Solar Neutrino Experiments Imply New Physics?
Phys. Rev. D **47**, 1298–1301 (1993)

334. (with G.E. Brown) A Scenario for a Large Number of Low Mass Black Holes in the Galaxy
Astrophys. J. **423**, 659–664 (1993)

335. My Experience in Teaching Physics
Am. J. Phys. **61**, 972–973 (1993)

336. (with G.E. Brown, A.D. Jackson and P.M. Pizzochero) The Hadron to Quark–Gluon Transition in Relativistic Heavy Ion Collisions
Nucl. Phys. A **560**, 1035–1074 (1993)

337. Possible Explanation of the Solar-Neutrino Puzzle
in: J.N. Bahcall, R. Davis Jr., P. Parker, A. Smirnov, and R. Ulrich (eds.), *Solar Neutrinos. The First Thirty Years*, Frontiers in Physics, Addison–Wesley, Reading, MA, 1994, pp. 65–68

338. Mechanism of Supernovae: The Bakerian Lecture, 1993
Philos. Trans. R. Soc. London A **346**, 251–258 (1994)

339. (with R. Bingham, J.M. Dawson and J.J. Su) Collective Interactions between Neutrinos and Dense Plasmas
Phys. Lett. A **193**, 279–284 (1994)

340. Theory of Neutrinos from the Sun
in: G. Ekspong (ed.), *The Oskar Klein Memorial Lectures, Vol. 2*, World Scientific, Singapore, 1994, pp. 1–15

341. Supernova Theory
in: G. Ekspong (ed.), *The Oskar Klein Memorial Lectures, Vol. 2*, World Scientific, Singapore, 1994, pp. 17–26

342. (with G.E. Brown) A Scenario for a Large Number of Low Mass Black Holes in the Galaxy
Astrophys. J. **423**, 659–664 (1994)

343. Stalin and the Bomb — The Soviet Union and Atomic Energy 1939–1956
Nature **372**, 281–283 (1994)

344. (with K. Gottfried and R.Z. Sagdeev) Did Bohr Share Nuclear Secrets?
Sci. Am. **272 (5)**, 84–90 (1995)

345. The Supernova Shock
Astrophys. J. **449**, 714–726 (1995)

346. (with J.J. Su, R. Bingham and J.M. Dawson) Nonlinear Neutrino Plasmas Interactions
Phys. Scripta T **52**, 132–134 (1995)

347. (with G.E. Brown) Observational Constraints on the Maximum Neutron Star Mass
Astrophys. J. **445**, L129–L132 (1995)

348. Obituary — Chandrasekhar, Subramanyan (1910–1995)
Nature **377**, 484 (1995)

349. (with R. Bingham, J.M. Dawson and J.J. Su) Anomalous Scattering and Absorption of Neutrinos in Dense Plasmas
in: P.H. Sakanaka and M. Tendler, *International Conference on Plasma Physics*, ICPP 1994, American Institute of Physics, Woodbury, NY, 1995, pp. 406–415

350. Cease and Desist
Bull. Atomic Scientists **51 (6)**, 3 (1995)

351. (with K. Gottfried and R.Z. Sagdeev) Nuclear Intrigues — A Reply
Sci. Am. **273**, 10 (1995)

352. Dark Sun: The Making of the Hydrogen Bomb
Science **269**, 1455–1457 (1995)

353. The Physical Review: The First Hundred Years
Phys. Today **44 (11)**, 77 (1995)

354. The Supernova Shock VI
Astrophys. J. **473**, 343–346 (1996)

355. Supernova Theory
Nucl. Phys. A **606**, 95–117 (1996)

356. Breakout of the Supernova Shock
Astrophys. J. **469**, 737–739 (1996)

357. (with R. Bingham, J.M. Dawson, P.K. Shukla and J.J. Su) Nonlinear Scattering of Neutrinos by Plasma Waves: A Ponderomotive Force Description
Phys. Lett. A **220**, 107–110 (1996)

358. Supernova Shock VIII
Astrophys. J. **490**, 765–771 (1997)

359. Influence of Gamow on Early Astrophysics and on Early Acceleration in Nuclear Physics
in: E. Harper, W.C. Parke, and D. Anderson (eds.), *George Gamow Symposium*, ASP Conference Series, Vol. 129, ASP, San Francisco, 1997, pp. 44–48

360. (with P.K. Shukla, R. Bingham, J.M. Dawson and L. Stenflo) Nonlinear Coupling between Intense Neutrino Fluxes and Dense Magnetoplasmas
Phys. Scripta **55**, 96–98 (1997)

361. (with P.K. Shukla, L. Stenflo, R. Bingham, J.M. Dawson and J.T. Mendonça) Nonlinear Generation of Radiation by Intense Neutrino Fluxes in Super-Dense Magnetized Plasmas
Phys. Lett. A **224**, 239–242 (1997)

362. (with P.K. Shukla, L. Stenflo, R. Bingham, J.M. Dawson and J.T. Mendonça) Generation of Magnetic Fields by Nonuniform Neutrino Beams
Phys. Lett. A **233**, 181–183 (1997)

363. (with P.K. Shukla, L. Stenflo, R. Bingham, J.M. Dawson and J.T. Mendonça) Nonlinear Propagation of Neutrinos in a Strongly Magnetized Medium
Phys. Lett. A **230**, 353–357 (1997)

364. (with P.K. Shukla, L. Stenflo, R. Bingham, J.M. Dawson and J.T. Mendonça) Stimulated Scattering of Neutrinos by Electron-Cyclotron-Harmonic and Convection Modes in Magnetoplasmas
Phys. Lett. A **226**, 375–377 (1997)

365. (with A.G. Petschek and G.I. Bell) J. Carson Mark — Obituary
Phys. Today **50 (10)**, 124 (1997)

366. (with V.N. Tsytovich, R. Bingham and J.M. Dawson) Collective Neutrino Plasma Interactions
Astropart. Phys. **8**, 297–307 (1998)

367. (with G.E. Brown) Evolution of Binary Compact Objects That Merge
Astrophys. J. **506**, 780–789 (1998)

368. (with P.K. Shukla, L. Stenflo, R. Bingham, J.M. Dawson and J.T. Mendonça) The Neutrino Electron Accelerator
Phys. Plasmas **5**, 1–3 (1998)

369. The Need for Nuclear Power
Environment and Nuclear Energy (1998) 39–41

370. (with G.E. Brown) Contribution of High-Mass Black Holes to Mergers of Compact Binaries
Astrophys. J. **517**, 318–327 (1999)

371. (with G.E. Brown and C.H. Lee) The Formation of High-Mass Black Holes in Low-Mass X-ray Binaries
New Astronomy **4**, 313–323 (1999)

372. (with P.K. Shukla, L.O. Silva, R. Bingham, J.M. Dawson, L. Stenflo, J.T. Mendonça and S. Dalhed) The Physics of Collective Neutrino Plasma Interactions
Plasma Phys. Controll. Fusion **41**, A699–A707 (1999)

373. (with P.K. Shukla, L. Stenflo, R. Bingham, J.M. Dawson, L.O. Silva and J.T. Mendonça) Neutrinos as Building Blocks of the Universe
in: J. Buechner, S.I. Axford, E. Marsch, and V. Vasyliunas (eds.), *Plasma Astrophysics and Space Physics*, Kluwer Academic Publishers, Dordrecht, 1999, pp. 731–737

374. The Treaty Betrayed
New York Review of Books **46 (18)**, 6 (1999)

375. Special Issue in Honor of the Centenary of the American Physical Society — March 1999
Rev. Mod. Phys. **71**, U6 (1999)

376. Quantum Theory
Rev. Mod. Phys. **71**, S1–S5 (1999)

377. Nuclear Physics
Rev. Mod. Phys. **71**, S6–S15 (1999)

378. (with G.E. Brown, C.H. Lee and H.K. Lee) Is Nova Sco 1994 (GRO 1655-40) a Relic of a GRB?
in: R.M. Kippen, R.S. Malozzi, and G.J. Fishman (eds.), *Gamma-Ray Bursts*, AIP Conference Series 526, AIP, Melville, New York, 2000, p. 628ff

379. The German Uranium Project
Phys. Today **53 (7)**, 34–36 (2000)

380. (with R. Christy) Oppies Colleagues Confirm his Leadership in Manhattan Project
Phys. Today **53**, 15 (2000)

381. (with G.E. Brown and C.H. Lee) Hypercritical Advection-dominated Accretion Flow
Astrophys. J. **541**, 918–923 (2000)

382. (with G.E. Brown, C.H. Lee and R.A.M.J. Wijers) Evolution of Black Holes in the Galaxy
Phys. Rep. **333**, 471–504 (2000)

383. (with G.E. Brown, C.H. Lee, R.A.M.J. Wijers, H.K. Lee and G. Israelian) A Theory of Gamma-Ray Bursts
New Astronomy **5**, 191–210 (2000)

384. Memoirs: A Twentieth-Century Journey in Science and Politics by Edward Teller with Judith Shoolery
Physics Today **54**, 11, 55–56 (2001)

385. (with G.E. Brown and C.H. Lee) High Mass Black Holes in Soft X-Ray Transients: Gap in Black Hole Masses?
in: Ch. Kouveliotou, J. Ventura, and E. van den Heuvel (eds.), *The Neutron Star–Black Hole Connection*, Kluwer Academic Publishers, Dordrecht, 2001, pp. 343–348

386. The Magic Furnace: The Search for the Origin of Atoms
Nature **411**, 888 (2001)

387. (with G.E. Brown, A. Heger, N. Langer, C.H. Lee and S. Wellstein) Formation of High Mass X-Ray Black Hole Binaries
New Astronomy **6**, 457–470 (2001)

388. (with G.E. Brown, C.H. Lee and S.F. Portegies Zwart) Evolution of Neutron-Star, Carbon-Oxygen White-Dwarf Binaries
Astrophys. J. **547**, 345–354 (2001)

389. (with G.C. McLaughlin, R.A.M.J. Wijers and G.E. Brown) Broad and Shifted Iron-Group Emission Lines in Gamma-Ray Bursts as Tests of the Hypernova Scenario
Astrophys. J. **567**, 454–462 (2002)

390. Enrico Fermi in Rome, 1931–1932
Phys. Today **55 (6)**, 28–29 (2002)

391. My Life in Astrophysics
Annu. Rev. Astron. Astrophys. **41**, 1–14 (2003)

392. (with N.D. Mermin) A Conversation about Solid-State Physics
Phys. Today **55 (6)**, 53–56 (2004)

393. Los Alamos and Cornell
Talk at a symposium honouring Robert Wilson
Fermilab Goldenbook http://history.fnal.gov/GoldenBooks/gb_bethe.html